Cloud Computing Security - Concepts and Practice

Edited by Dinesh G. Harkut

Published in London, United Kingdom

IntechOpen

Supporting open minds since 2005

Cloud Computing Security - Concepts and Practice
http://dx.doi.org/10.5772/intechopen.83221
Edited by Dinesh G. Harkut

Contributors
Akashdeep Bhardwaj, Sam Goundar, Georgios Karakonstantis, Charles Gillan, Laud Ochei, Yousef Ibrahim Daradkeh, Petr M. Korolev, Victor Telnov, Yuri Korovin, Dr. Dinesh G. Harkut, Svetlana Aristova

First published in London, United Kingdom, 2020 by IntechOpen
IntechOpen is the global imprint of INTECHOPEN LIMITED, registered in England and Wales, registration number: 11086078, 7th floor, 10 Lower Thames Street, London, EC3R 6AF, United Kingdom
Printed in Croatia

British Library Cataloguing-in-Publication Data
A catalogue record for this book is available from the British Library

Additional hard and PDF copies can be obtained from orders@intechopen.com

Cloud Computing Security - Concepts and Practice
Edited by Dinesh G. Harkut
p. cm.
Print ISBN 978-1-83880-702-3
Online ISBN 978-1-83880-703-0
eBook (PDF) ISBN 978-1-83880-704-7

For EU product safety concerns:
IN TECH d.o.o., Prolaz Marije Krucifikse Kozulić 3, 51000 Rijeka, Croatia, info@intechopen.com or visit our website at intechopen.com.

Meet the editor

Dr. Dinesh G Harkut is currently working as an Associate Professor at PRMCEAM, Badnera, India in the Computers Science & Engineering Department. He obtained his Bachelors, Masters of Engineering Degree (CSE), and Ph.D. (CSE) from SGBAU Amravati University, Maharashtra, India. He also obtained his Masters Degree in Business Administration and Ph.D. (Business Administration). His primary research interests are in computer AI, big data, analytics, embedded systems, and e-commerce. He has supervised around 18 Master degree and 24 Bachelor degree students. He has published 47 papers in refereed journals and published 6 books with international publishers. He has filed 2 patents and published in his name in India and organized various workshops, sessions, conferences, and trainings. He is a Member of the Board of Studies (Computer Science & Engineering) and Recognized PhD Supervisor at SGBAU Amravati University, Maharashtra, India. He holds membership with various professional bodies in different capacities: Fellow member of IETE, New Delhi; Life member of ISTE, New Delhi; Senior Member of UACEE, USA; Senior Member of IEDRC, HK; Professional member of IAENG, Hong Kong and Member of European Alliance for Innovation, Belgium.

Contents

Preface

As the world is increasingly interconnected,
everyone shares the responsibility of Securing Cyberspace and
Security is always excessive until it is not enough.

- Anonymous

This book is intended to teach you about the most important security controls for your most important assets quickly and correctly, whether you're a security professional who is somewhat new to the cloud, or an architect or developer with security responsibilities.

Cloud computing is the third wave of the digital revolution and it is actually a spectrum of things complementing one another, building on a foundation of sharing. Cloud computing enables simplified functionality of platforms and infrastructure used in IT-enabled industries, so that the end-users can consume what and when they want and pay only for the service they use. Cloud computing can be characterized as Internet-based computing in which many remote networked servers facilitate shared data-processing ventures, centralized storage, and access to services or resources online. With the advent of business process outsourcing of IT and IT-enabled services, cloud computing has gained significant commercial interest. Inherent dualities in the cloud computing phenomenon are spawning divergent strategies for cloud computing success. The public cloud, hybrid clouds, and private clouds now dot the landscape of IT based solutions. Cloud computing allows for both large and small organizations to have the opportunity to use Internet-based services so that they can reduce start-up costs, lower capital expenditures, use services on a pay-as-you-use basis, access applications only as needed, and quickly reduce or increase capacities. However, these benefits are accompanied by a myriad of security issues, and this valuable book tackles the most common security challenges that cloud computing faces. In almost all organizations, security has to fight for time and funding, and it often takes a back seat to implementing features and functions. While many of the security controls and principles are similar in cloud and on-premises environments, there are some important practical differences. As the title states, this book is a comprehensive guide to secure your cloud environments.

I would like to convey our appreciation to all contributors including the accepted chapters' authors. I owe special thanks to Ms. Mia Vulovic, Author Service Manager and Ms. Klara Mestrovic, Commissioning Editor, IntechOpen, London, UK for their kind support and great efforts in bringing the book to fruition. In addition, I also appreciate all those who worked in the background and have assisted in formatting the book.

Dr. Dinesh G. Harkut
Dean and Associate Professor,
Department of Computer Science and Engineering,
Prof. Ram Mehge College of Engineering and Management,
Badnera-Amravati, Maharashtra, India

Chapter 1

Introductory Chapter: Cloud Computing Security Challenges

Dinesh G. Harkut

1. Introduction

Cloud Computing is currently one of the hottest topics in computing and information technology (IT). The term "Cloud Computing" does not represent a host of new technologies, rather these technologies are combined and effectively upgraded so that they enable new IT services and new business models.

Cloud computing is a technology paradigm that is offering useful services to consumers. Cloud Computing has the long-term potential to change the way information technology is provided and used. The entire cloud ecosystem consists of majorly four different entities which plays vital role to fulfill the requirements of all the stake holders. The role played by each individual depends on their position in the market and their business strategy. These most prominent entities in the cloud ecosystem are:

- **Cloud Service Provider:** it provides cloud services available to cater the needs of different users from different domain by acquiring and managing the computing resources both hardware and software and arranging for networked access to the cloud customers.

- **Cloud Integrator:** the facilitators, one who identify, customize and integrate the cloud services as per the requirement and in accordance with the customers' needs. It plays the important role of matchmaking and negotiating the relationship between the consumer and producer of the services.

- **Cloud Carrier:** it is an intermediary which facilitates the connectivity and takes the cloud services at the doorsteps of end-user by providing access through different network access and devices.

- **Cloud Customer:** the actual user of services extended by the service provider which may be an individuals or organizations which in turn may have their own end-users like employees or other customers.

2. Types of service models

Cloud service providers harness the benefit of huge computing resources span over large geographical area to provide seamless, efficient and reliable services to customers at marginal price. The computing resource deployed over the Internet comprises hardware and application software and OS used in virtualization, storage

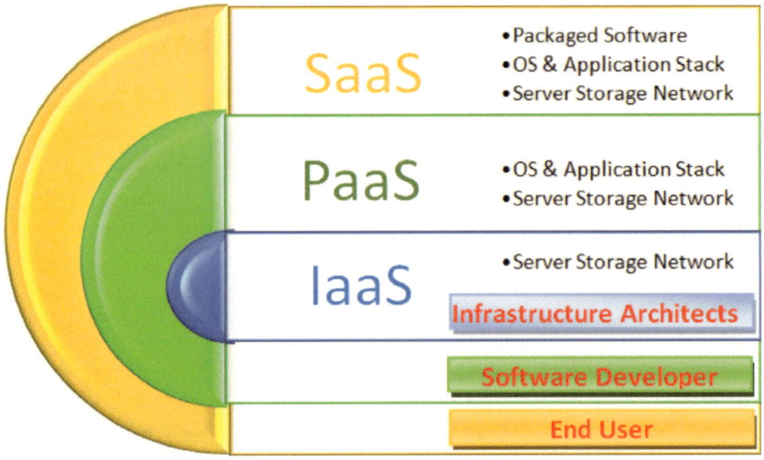

Figure 1.
Cloud service model.

and compute purposes. There are basically three different service models (**Figure 1**) of offering high-volume low-cost services to the end user:

- Software as a Service (SaaS)
- Platform as a Service (PaaS)
- Infrastructure as a Service (IaaS)

2.1 Software as a service (SaaS)

In this model, various applications are hosted by a cloud service provider and publicized to the customers over internet, wherein end user can access the software using thin client through web browsers. Here all the software and relevant data are hosted centrally on the cloud server. CRM, Office Suite, Email, Games, Contact Data Management, Financial Accounting, Text Processing etc. are typically falls under this category.

2.2 Platform as a service (PaaS)

A PaaS is typically is a programming platform for developers. This platform facilitates the ecosystem for the programmers/developers to create, test, run and manage the applications. It thus provides the access to the runtime environment for application development and deployment tools. Here developer does not have any access to underlying layers of OS and Hardware, but simply can run and deploy their own applications. Microsoft Azure, Salesforce and Google App Engine are some of the typical examples of PaaS.

2.3 Infrastructure as a service (IaaS)

IaaS facilitates availability of the IT resources such as server, processing power, data storage and networks as an on demand service. Here user of this service can dynamically choose a CPU, memory storage configuration according to needs. A cloud user buys these virtualized and standardized services as and when required.

For example, a cloud customer can rent server time, working memory and data storage and have an operating system run on top with applications of their own choice.

3. Types of deployments

Furthermore, these services can be deployed into Public Clouds, Private Clouds or Hybrid Clouds; each has its own advantages and disadvantages.

3.1 Public cloud

In the Public Cloud delivery mode, all the physical infrastructure are owned by the provider of the services which were provided off-site over the Internet hosted at cloud vendor's premises. Here the customer has no control and limited visibility over where the service is hosted as all these massive hardware installations are distributed throughout the country or across the globe seamlessly. This massive size enables economies of scale that permit maximum scalability to meet varying requirements of different customers and thus provides greatest level of efficiency, maximum reliability through shared resources but with rider cost of added vulnerability.

3.2 Private cloud

In case of Private Cloud mode, entire infrastructure is owned, managed and operated exclusively by the organization or by a third-party vendor or both together and is hosted on the organization premise using virtualization layer. It also facilitates flexibility, scalability, provisioning, automation and monitoring and thus offers the greatest level of control, configurability support, high availability or fault tolerant solutions and advanced security which is missing in public cloud. Basically, very concept of private clouds is driven by concerns around security and keeping assets within the firewall which results it to significantly more expensive with typically modest economies of scale.

3.3 Hybrid cloud

As name suggest, Hybrid Cloud includes a variety of product mix from both Public and Private Cloud options sourced from multiple providers at added cost to keep track of multiple different security platforms by ensuring all aspects of business to communicate with each other seamlessly. In case of Hybrid approach, operational flexibility, scalability, efficiency and security are properly balanced by hosting mission critical applications and sensitive data protected on the Private Cloud and generic application development, big data operations on non-sensitive data and testing on the Public Cloud. Hybrid Cloud thus leverage benefits of both Public and Private Cloud by maintain balance between the efficiency, cost saving, security, privacy, and control.

The combination of the different service and deployment models enables different business models with new business roles. A cloud service is likely to have many layers of abstraction that build on top of each other with define roles and duties. Accessibilities of these predefine services to the end user depends on the different service model. Abstraction layers of standard Cloud model is depicted in adjoining **Figure 2**. Service providers may adapt and compose several services into one, which is then offered to the cloud customers.

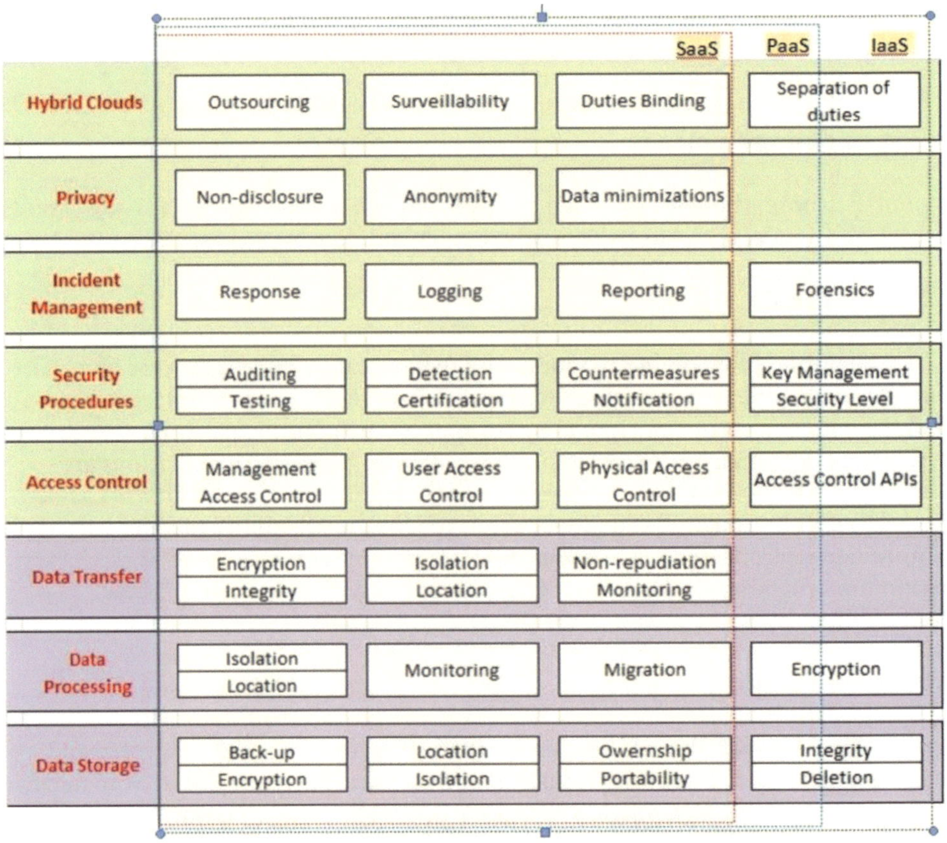

Figure 2.
Abstraction layers of model cloud.

Cloud computing has emerged as a major shift in how computing resources are deployed and consumed both by individuals and enterprises Cloud computing is an approach that covers a wide spectrum of cloud tools and models. This technology has a lot of potential and promises its consumer an enhancement in agility, efficiency and profitability by offering software, platform, and infrastructure delivered as services at very negligible cost by reducing up-front investment and ease of use by providing most user and eco-friendly operations. Like other technology, cloud also offers many benefits which come with some rider cost associated with it. Cloud too has its weaknesses and that is security.

Essentially, security in the cloud environment does not differ from the one in the traditional computing model. In both cases, the major focus is on the issues of protecting data from theft, leakage or deletion. Unlike in traditional computing model, issue of security in the cloud is slightly different. When individual users or organizations move computer systems and data to the cloud, security responsibilities become shared between user and cloud service provider. When an increasing number of individual users and businesses are moving their precious data and entire IT infrastructures to the cloud, it is natural to start wondering how security and privacy are handled in the cloud.

Due to its intrinsic nature, however, the cloud environment highly susceptible to security threats as compared to its counterpart as data is stored with some third-party provider and accessed on the web which increases the overall vulnerability and thus affects overall reliability. Moreover, as most of the precious data is transferred to the cloud, it is difficult to maintain its integrity and thus overall data security is compromised.

Furthermore, with the advancement of technology and passage of time, the entire cloud ecosystem has evolved and instead of relying on single cloud provider for buying/renting a cloud service, individual user or business organizations having freedom and flexibility to exploring more options to select multiple service providers simultaneously for different needs from pool of different cloud service provider and thus eventually leads to more diffuse, seamless integrations of multiple service providers term as fog computing. To make thinks further complicated, data and services may be replicated horizontally among these multiple service providers and as a consequence, it is often extremely difficult to determine the physical location as to where the data is being stored or processed at any one time.

All this constituted the obvious security implications as data is transmitted and stored in different locations over the Internet and shared among multiple service providers simultaneously. Such data is neither within the control of the individual/owner nor the individual service provider specifically in fog environment which is common now a days. Apart from just data, virtualization and applications are equally important security issues in cloud computing. Thus, Security has severe impact on the overall decision making process as to whether an individual or organization will adopt for the cloud services or not.

Though cloud services have ushered in a new age of transmitting and storing data and cloud has its own beneficial power but it is imperative to take focused security approach, reviews the changes needs to undertake before making decision to migrate to the cloud. Some of the key aspects of cloud security in nut shell are depicted in **Figure 3**.

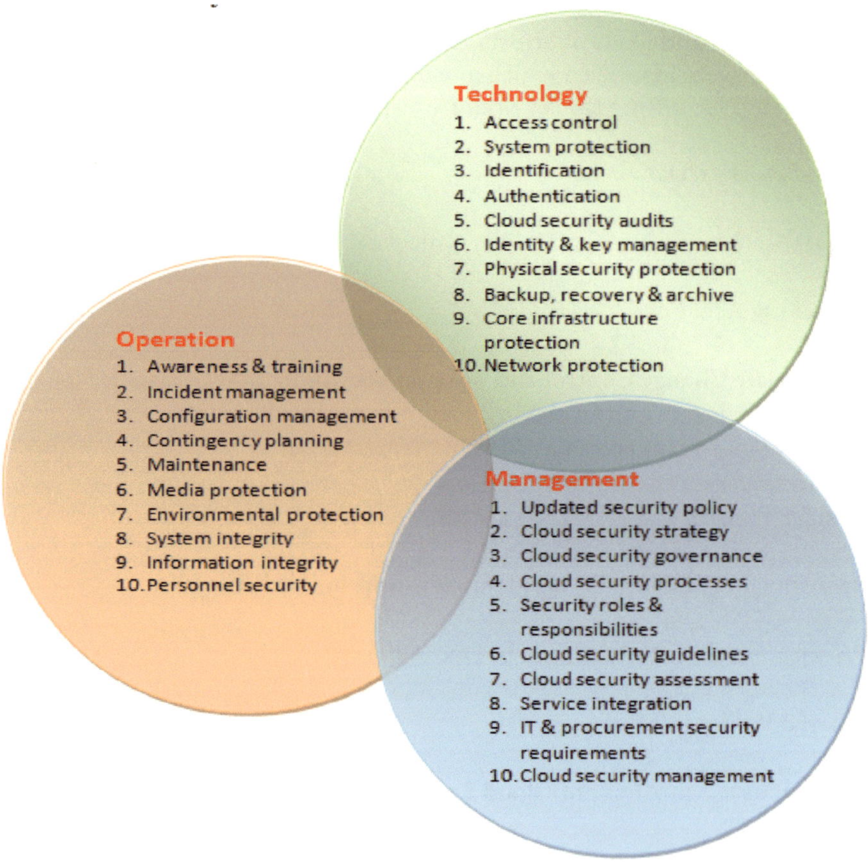

Figure 3.
Aspects of cloud security.

Cloud Security Simplified:

- Access Control
- System Protection
- Personal Security
- Information Integrity
- Cloud Security Management
- Network Protection
- Identity Management

4. Vulnerabilities and threats

Cloud computing being a modern technology offers numerous advantages. In order to harness all these benefits, one has to scrupulously investigate as many cloud security measures as possible. These concerns may vary from vulnerability to malicious code penetration to hijacked accounts to full-scale data breaches. Based on literature searches and analysis efforts, some of the major cloud-unique vulnerabilities and threats were identified which one must consider before making decision to migrate to cloud for opting the services are as follows:

1. Data Breaches/Data Loss
2. Denial of Service Attacks/Malware Injection
3. Hijacking Account
4. Inadequate Change Control and Misconfiguration
5. Insecure Interfaces and Poor APIs implementation
6. Insider Threats
7. Insufficient Credentials and Identity/Compromised accounts
8. Weak control plane/Insufficient Due Diligence
9. Shared Vulnerabilities
10. Nefarious use or Abuse of Cloud Services
11. Lack of cloud security strategy/Regulatory violations
12. Limited cloud usage visibility

4.1 Data breaches/data loss

Cloud computing and services being relatively new and enable accessing remote data via the Internet is the most vulnerable source for misconfiguration or exploitation. This very intrinsic property of cloud becomes unique set of characteristics which make it more vulnerable to all form of data breaches. Data breaches or losses can be any form of cyber security attack in which confidential or sensitive information is stolen, viewed or used by an unauthorized stranger or it may the result out of accidental deletion by service provider or a natural catastrophe, like fire outbreak or earthquake. This may results to the loss of intellectual property (IP) to rivals, impacts the competitive edges, financial losses out of regulatory implications, affecting brand value and goodwill of organization and thus overall market value may be at stake as it foster mistrust from customers and business partners. Though Encryption techniques can protect data but at the cost of system performance. Thus robust and well-tested Data breach avoidance, data loss preventions, data backup and recovery data management strategy must be adopted before making up mind to migrate to cloud.

4.2 Denial of service attacks/malware injection

The basic framework of cloud which offers scalability and speed also becomes nurturing ground for delivering super scalable malware. Cloud applications themselves are great weapon for spreading the malicious attacks on a large scale to cause greater harm like hijacking accounts, breaching data. Malware injections are basically code scripts which are embedded into the basic cloud service modules thus run as legitimate instance having access to all the sensitive resources and thus intruder can eavesdrop, compromise the overall integrity of vital information. Denial of Service attack (DoS) makes valuable services unavailable to the legitimate user thus hamper the overall performance and security. DoS may act as catalyst and used as smokescreen to hide the malicious activities bypassing the firewall of cloud and thus can spread easily to cause greater harm instead of infecting one device.

4.3 Hijacking account

The recent growth and easy adaption of cloud services by organization leads to altogether new set of issues related to hijacking account. Imposter now can easily exploit the ability to gain access to login credentials and thus the sensitive data comprises of business logic, functions, data and applications stored on the remote cloud. Account hijacking which includes scripting bugs, reused password, cross-site scripting enables the intruder to falsify and manipulate information. Man-In-Cloud Attack, Key-logging, Phishing, and buffer overflow are some other similar threats which eventually leads to theft of user token which cloud platform uses to verify each individuals without requiring login credentials typically during data updation or sync. The impact of the account hijacking can be severe, some even leads to significant disruption of business operations by means of complete eliminations of assets and capabilities. Thus account hijacking needs to be dealt seriously as tangible and intangible impact out of leakage of sensitive and personal data may damage the reputation and band value.

4.4 Inadequate change control and misconfiguration

Volume and scope of the various resources used in cloud environment augmented with complexity and dynamism of resources poses major challenge in configuring effectively for efficient use. Inappropriately configure precious computing resources, results in making these resources soft target for vulnerable malicious undesired activities and thus entire cloud repositories may exposed to intruders. The overall business impact depends on the nature of the misconfiguration, and how quickly it has been detected and resolved. Excessive undesirable permission, unrestricted access to ports and services, unsecured data storage, unchanged default credentials & configuration settings, disabling standard security controls, logging & monitoring are some typical issues related with misconfiguration which must be dealt with utmost care by continuously scanning for misconfigured resources in real time as traditional change control and configuration management technique becomes ineffective in cloud environment.

4.5 Insecure interfaces and poor APIs implementation

Application Programming Interfaces (APIs) as name suggests is an interface between the system and outside un-trusted entities most exposed parts of a system accessible via the Internet, facilitates users to customize their cloud experience and also indirectly provide the safe conduit or entry points for strangers. A poorly designed weak set of interfaces exposes organizations precious sensitive resources to various security issues related to confidentiality, integrity, availability, and accountability. Apart from giving programmers the tools to build and integrate their applications with other job-critical software, API also serves to authenticate, provide access, and effect encryption. The cloud assets can be compromised if the vulnerability of an API which lies in the communication that takes place between applications is exploited. Thus standard and open API frameworks must be referred while designing the interfaces which may help to protect against both accidental and malicious attempts to circumvent security.

4.6 Insider threats

The human intervention in data security has many faces and many sources. The insider human element may be from any hierarchy; both service provider and client organizations can abuse their authorized access to the organization's or cloud provider's networks, systems, and data as they are uniquely positioned to cause damage without even breaking the firewalls and other security defense mechanism. The human element of data security has many faces and being authorized and operated on a trusted level, these insiders may misuse information or perform nefarious activities through malicious intent, accidents, carelessness or malware. Various measures to mitigate the consequences of insider threats includes routine audits of on-offsite servers, frequent change in passwords, confined privileged access to security systems and central servers to limited numbers of employees apart from controlling access and offering business partnerships to the employees. Prevention is better than cure; dealing with such category of threat would become more expensive and complex as it involves containments, forensic investigation, escalation, surveillance and monitoring.

4.7 Insufficient credentials and identity/compromised accounts

Inadequate credential, identity or key management may leads to unauthorized access to data and information. As a result, malicious intruders camouflaged as genuine users can manipulate the sensitive data. If the impostor manages to gain access to cloud user's credentials, it can target the entire resource s of cloud along with the user organization's assets and even influence the organization's administrative user as well. Other tenants of the same cloud are also at high risk to security incidences and breaches. An Automated regular rotation of cryptographic keys and passwords, removal of unused credentials, implementation of proper scalable central programmatic credential management system, and use of multifactor authentication process are some of the measures which must be undertaken by the cloud provider to deviate the risk of data breaches. Moreover, due diligence should be taken to ensure that third parties to whom cloud provider may have outsources operations or maintenance work satisfy the requirements of security as contracted by cloud service provider because it indirectly levitate the threats and compromised the overall security. Strictest credential access, multifactor authentication, segregated and segmented accounts are some of the suggested measures one should opt for to mitigate the risk.

4.8 Weak control plane/insufficient due diligence

Non-standard data formats, non-standard APIs, and excessive reliance on loud provider's proprietary tools make it difficult and expensive affaires to migrate from one vendor to other. This may results in either cloud provider will start exploiting or in case if for some reason cloud provider ceases its operation and goes out of business, moving data to other in timely manners becomes hectic and eventually may result in loss of data too. Thus to avoid such grim situation of Vendor lock-in, adequate control plan and due diligence should be in place before making decision migrating to any cloud. Any hasty decision without anticipating the quality and nature of services from cloud provider may pose security risk, especially when the desired services are bound and control under legal and statutory obligations or services hired for handling highly sensitive or personal or financial data. Cloud service user must perform due diligence and ensure that proposed cloud service provider possesses an adequately strong control plane in place; absence of this could results in data loss, either by theft or corruption. Apart from technical issues discussed above, one equally important parameter which must be given due weightage in decision making process is people factor. If a person in charge is unable to exercise full control over data security, infrastructure and verification, then security, integrity and stability of data may be stake.

4.9 Shared vulnerabilities

Multi-tenancy feature of cloud makes cloud services cost effective for individual organization but incidentally it leads to yet another security issue. Exploitation of system and software vulnerabilities within cloud infrastructure, services results into failure to maintain physical and logical separation among different tenants in multi-tenant environment. This failure to maintain separation can further be exploited by intruders to gain un-authorized access from one tenant's resource to others. Such attacks can be accomplished by exploiting the vulnerabilities of either cloud provider or any of the tenants whose security is more vulnerable. This may results in increasing the attack surface, leading to an increased chance of data leakage. Moreover, the cloud security by default is a shared responsibility of both cloud service provider and client organization, so proper understanding is imperative to implements effective security. Failure to achieve this seamless integration for security implementation can result in data and resources being compromised.

4.10 Nefarious use or abuse of cloud services

Intruders by exploiting the vulnerabilities of cloud computing resources may target user's cloud provider's resources to host malware activities. Intruder either may launching DoS attacks and thus makes services unavailable to legitimate users or these resources can be used for some illegitimate use for illicit purpose like mining crypto-currency, automated click trailing, brute-force attacks for security breach by intruders and while the customer foots the bill. The bill could be substantially high as activities like mining requires huge resources. Attackers may use the clouds exceptional storage capacity to store and propagate malware and illicit activities like sharing of pirated software, books, videos or music and invites legal consequences in intellectual copyright fines and settlements which can be even more cost prohibitive. Furthermore, complexity of cloud service implementation aids intruders to hide and remain undercover for prolonged period of time and such unnoticed threats, risks and vulnerabilities poses more challenges for legitimate service provider and user. To restrain the nefarious use and abuse of cloud services and mitigate the risks posed by cloud service usage one must have to procuring security technology for actively monitoring cloud infrastructure usage and devise proper security guidelines which define what are the legitimate and appropriate behavior and what leads to abuses and methods of detecting such behaviors.

4.11 Lack of cloud security strategy/regulatory violations

It is imperative to formulate a strong cloud security strategy, regulations and risk management policy should be devise before making mind to migrating to cloud provider for various services instead of simply lift and shift without any due diligence. Mostly many organizations are bound by and force to comply with certain rules, regulations and law of land of origin and these compliances should be center point for overall security policy. Sensitive health data, private student data, personal financial data, proprietary intellectual property data, research data and confidential business logics constitutes different category of data which are typically migrated to cloud for various services and mostly protections of these data are cover under respective apex authorities or commission and infringement of any kind will invite the formidable fine and penalties. Security architectures and framework must be aligns with the underlying business goals and objectives. Cloud provider being third party, upon agreeing to provide the services, also become liable for extending the appropriate security measures as weak security can lead to financial loss, reputational damage, legal repercussions, and fines.

4.12 Limited cloud usage visibility

The moment organization decides to migrate the assets and operation to the cloud, it starts losing the overall visibility and control over those assets. The ability to decide, visualize and analyze whether the services offered by clouds are safe or malicious, decides the degree of visibility of cloud usage. Even though organizations are hiring the services of cloud provider, still it is imperative on their part to perform analysis and run time monitoring. To enhance the cloud visibility and thus mitigate the risk, it is crucial to develop comprehensive solution that brings people, process and technology at one common place and elucidate accepted cloud usage policies to each and every stack holder. Otherwise lack of awareness about organizations governance controls and policies may results in placing sensitive data in public access and compromising the cloud containers by inappropriate setup of cloud services. Thus lack of governance, lack of security and lack of awareness leads to catastrophic risk. Installing firewall, implementing organization wide zero-trust model, run time analysis of outbound activities and keeping track of anomalies are some of the measures which will be helpful in restraining the suspicious behaviors and mitigating the overall risk.

5. Conclusion

Cloud is new buzzword and evolving at a phenomenal speed, even in the context of the fast-moving IT sector and becoming increasingly in demand around the world. As it evolves, lack of faith in the security features imparted by cloud is cited as main barriers and concerns which discourage users putting their confidential data into this faceless nebulous and intangible entity known as the cloud. Information security and data protection are the two main concerns which stand in the way of a wider deployment and acceptance of cloud. Over a passage of time, most powerful security standards are emerging and constantly evolving aiming to overcome many of these challenges. Clearly, there are both challenges and opportunities with the cloud and due to the economics of scale, a cloud provider are opting for a dedicated team of security specialists and cloud data centers have physical protection on par with military installations thus able to provide vastly better security procedures, physical protection than any small or medium-sized enterprise. In summary, as with each new technology, Cloud is a double-edge sword and clearly there are both challenges and opportunities with the cloud.

Author details

Dinesh G. Harkut
Prof. Ram Meghe College of Engineering and Management, Sant Gadge Baba Amravati University, India

*Address all correspondence to: dg.harkut@gmail.com

Chapter 2

Cloud Computing Security Services to Mitigate DDoS Attacks

Akashdeep Bhardwaj and Sam Goundar

Abstract

This chapter focuses on the challenges and risks faced in cloud security services in the areas which include identity access management, web security, email security, network security, encryption, information security, intrusion management, and disaster management while implementing a cloud service infrastructure. This chapter endorses the best practices in successfully deploying a secure private cloud infrastructure with security measures and mitigation and proposed a unique three-tier infrastructure design to mitigate distributed denial of service attacks on cloud infrastructures.

Keywords: threats/vulnerabilities, security policies, data protection/security, firewall, security model, monitor traffic, authorization

1. Introduction

Cloud computing is vastly increasing in demand for its popularity. Cloud services deliver up to its expectations if properly maintained. Users choose cloud because the cost is affordable, is easy to access, and has a positive uptime. Unfortunately, a high number of cloud users face difficulties when issues arise such as the frightening news about data confidentiality and integrity which gets posted online all the time, and they are in darkness when such situations occur [1]. In this modern age, cloud computing has been progressively popular within the IT organizations, and we notice many institutes are moving most of their IT services towards the cloud services here in Fiji to improve their information communication technology (ICT) service delivery to the clients or stakeholders. It is important for any organization to do an appropriate background research before making decisions of upgrading for which type of cloud services they are acquiring, depending on the organization's requirements. Many organizations prefer a private cloud infrastructure which has many advantages compared to the public cloud and hybrid cloud services; however, administrators tend to overlook that private cloud infrastructure comes with an exceptional set of challenges and risks. Cloud computing security service categories are identified and illustrated as follows:

- Identity access management
- Data loss prevention
- Web security

- Email security
- Network security
- Encryption
- Information security

2. Literature review

Cloud computing security keeps on changing as new technologies emerge. Services provided by the three basic cloud service models, which are infrastructure as a service (IaaS), platform as a service (PaaS), and software as a service (SaaS), give more outbreak to exploit such state-of-the-art models. As the cloud expands, so does its vulnerabilities [2]. It is the hosting providers or administrators duty to ensure that these vulnerabilities are kept on patching up as new threats arise. One must always be on the look for any threats coming their way towards the cloud servers. If such threats enter the cloud, it could be devastating for the cloud hosting providers and even the cloud servers itself. We as human beings need to keep in mind that a person who wants to get the data for their benefit or even for fun purposes [3] creates those threats. There are certain software programmed and integrated to the cloud server to automatically mitigate a certain level of threat, and an example is a web application firewall (Barracuda).

Stated in the research paper regarding a study on security model in cloud computing, we vastly agree to the statements stated such as the security being a real-time obstacle of the everlasting picture and foundation of cloud computing [4]. Furthermore, this research will now move towards focusing on the security aspect and its services shown by cloud computing itself as keeping in mind the increasing need for security in the cloud as we see a new day moving forward in our daily lives.

This paper starts with cloud identity access management as the first level of cloud computing security service that we identified based on various researches conducted. Whenever a user has established the connection to the cloud, the user will need to login and access the cloud resources in order to drive forward the idea of hosting applications, websites or even doing online sales securely through a secure login tunnel. This has to be fully done by successful authentication and authorization to avoid loss of data and identity being manipulated which could lead to unwanted access to the cloud system [5].

The need to have identified information in this case, which related to identity access management itself, first needs to be synchronized so that there are no conflicts when identifying the cloud user [6]. One needs to keep in mind there are many users who have the same name although their username can differentiate the users and their level of access. For user information to be synced properly, the old user data will need to be checked if there are any, which were used previously, and it should not match with the new data. Such scenarios occur when a user cancels their online subscription to a cloud host provider and comes back after a few years wanting to again host their applications on the cloud [7]. This reflects on mostly public clouds. This can also be hinted at a private cloud when the administrator permanently deletes the user's data, which in this case is the user's login and registration details to focus on. A hacker can pose as a new user and easily gain access to the cloud system if he/she is able to manipulate the registration and other details.

This can happen when the administrator is adding a new identity to the system, and if the administrator is not careful, the system does not identify old data. This will lead to an identity within the cloud system that will gain access to certain module level-based information because the identity has not been synced and verified with real time updates [8]. Another issue could be confidentiality, which focuses only on authorized access to cloud data; an authentication that is related to checking of the received data to be from a legit source and integrity, which relates to only authorized party, should be allowed to modify data in the cloud [9].

Building trust in cloud computing services may help prevent data loss to some extent, but it does not guarantee it. The cloud server needs to be equipped with state-of-the-art hardware and software in order to prevent such issue. This service protects data from being lost based on the rules deployed on cloud servers. Data can be lost in various ways such as the hacker sends a malicious file, which infects the cloud server and deleting the files and folders [10].

Data storage repository must be secured enough to handle such attacks although the level of attacks varies from high to low and each attack is to be considered no matter which level that threat is. Suppose a low level of threat occurs in the cloud server where that data is stored and the administrator does not take any action to fix the issue or just ignores it thinking that it is a small issue, it could multiply and do its job on the storage server leading to data loss [11].

Securing the storage in the cloud is very important where the storage or based in geographical location or not, but at the end of the day, the storage repository is linked to a network and that is enough information for an advanced hacker to easily delete the data by entering into the system from just a small script which will eventually grow to a virus or Trojan to inflict the damage.

If the administrators are not monitoring such scenarios, that virus can do the damage to the storage server. In such cases, it may send a lot of traffic request to the storage server, and this can result in overload. Such case is mostly described as denial of service attacks. With DoS attacks, the server will notice a change in traffic load coming in, and if there are no intelligent applications installed in the server to mitigate, there could be serious implications. These attacks can corrupt data, delete data, and data loss. This is a common issue faced by a lot of users, which ultimately will become the administrator's responsibility.

We noticed a virus spread across the globe called WannaCry, which is a ransomware virus where it locks down your computer system and asks for money in order to unlock the affected system. This type of attacks can lead to data loss as well. Supposedly, this threat can affect the data stored in the cloud server, which is definitely huge on a threat level. Microsoft had to realize patches for their operating systems in order to prevent such attacks. This results in a lot of distress around the globe and was one of most talked about attacks. It not only inflicts damage to the affected system, but it also has the ability to destroy the data itself, which is stored in any system [9]. One must be very careful of such attacks if not then data loss is inevitable. Such attacks are a wake-up call for cloud system, which does not have any type of data loss prevention techniques implemented, and if such techniques are implemented, then the administration must map out ways to block or to prevent data being lost. Therefore, security rules need to be in place to avoid customers from being frustrated with one of the major issues, which are data loss [4].

Web security plays a vital role as well in clouds. While the servers are hosted in clouds, websites and applications are also hosted in it which combines the functionality to work with cloud resources and deliver as expected to customers. Protection against virus and malware nowadays is very common as new types of such threats emerge almost every day. In cloud, all folders are synchronized at all times as the user updates their data. What could happen is that if a malware enters the cloud and data sync is taking place, the malware gets synched together, spreading around with the configured account, which is the source in which the malware entered into the system [11].

Hosting service providers for cloud-based will need to get a good web application firewall (WAF), which can prevent attacks to web servers and applications. Traffic going in and out of the web server needs to pass through WAF in order to check for malicious responses [12].

As proposed in the paper by Fernandez et al [13], web application scanner and a cloud-based web application firewall can be used to identify vulnerabilities and scan for sensitive data [13]. This type of scanner is very useful in a cloud computing environment. The cloud-based web application firewall will also be integrated with the scanner. The first step will be the scanning process followed by filtering unwanted request, keeping in mind that these unwanted requests are the virus and attacks coming into the system. In their paper, they have also stated that the firewall can control the web application communication via HTTP based on the rules for authorizing and with the main purpose for it to stop SQL injection, XSS, and other types of similar attacks on the cloud servers [12]. What our research has looked into is one of the WAF available for purchase called Barracuda. This application is very useful as it generates a whole lot of data that is not required for processing based on the traffic flow in which the attackers can come in and out of the system. This application has the ability to scan, put cloud applications and websites behind a state-of-the-art firewall system, and monitor traffic to name some of its core functionalities. When we look at a WAF system for cloud, we must have reports generated in order to do research that is more thorough from where the particular attack is coming from and how these attacks can be mitigated. The WAF provides a solution to every attack or vulnerability that is present in the generated report as well. This firewall will be able to stop unwanted traffic into the system, keep the cloud servers safe, and transfer those IP addresses that are suspicious to the suspicious list from the whitelist causing it to be classified as a threat [14]. The users can do online banking securely and other tasks that they would prefer to be done under a secure application layer.

Email security is being implemented in clouds as well. It has major advantages. Any inbound and outbound emails will need to go through email security protocols to ensure that the user sending and receiving the email does not contain any type of malicious data, which can affect the customers' online activity in any way. This could also lead to having a bad impact on web servers as well if proper security protocols are not in place to filter malicious emails. Cases of security policies need to be implemented in order to run the workflow of emails and filtering unwanted emails [15].

As outlined in the paper published as from Barracuda itself, using such application will not limit the functionalities of email security being deployed in the cloud servers. Some things to notice about the paper is that they have outlined the suit for the cloud services with the following combinations for the advanced package, multilayer security which extends the protection for the email also being integrated with Office 365 which is currently being well-known for its cooperative feature for an organization provided by Microsoft. Multilayer security is one of the core features that the email security giant company looks in depth, for the application itself is being a guard against threats arising from emails, data loss protection through spam emails, data being leaked with encryption, and all the email contents being inspected. Another advanced feature that they explained in their paper was cloud-based archiving. This feature is very important, and emails need to be archived frequently for an organization. Such feature in the cloud will enable users to retrieve any previous email at any time from any device, and this can be from any cloud environment whether it being the hybrid cloud-based environment, Microsoft 365, Microsoft Exchange, and even any other types of email service being used on-premise. They also mentioned retrieving emails such as cloud-based backup and recovery features. This feature will allow the administrators to retrieve any email from the frequent backup storage and send it to the live server so that the user requesting for the email can view and retrieve their contents for that particular email [16].

According to Rawezh Tara and Nashwans' paper based on private cloud and implementation of software, routing in it identifies the use of virtual private network (VPN), which enables the ability to ensure that the user who is logged into the cloud service can do their online transactions without any issues. The attackers will not be able to judge where or how the data is being transferred to. This creates a secure environment for customers doing online shopping or banking. It is a good practice to provide VPN to users who are already logged into cloud service. Each user will have a VPN client profile. Using this they can establish the VPN connection, and a secure tunnel is enabled, and authentication is being done on the data center firewall end [17].

Only two types of users use VPN tunnel, which will be the employees and the cloud administrator. The VPN tunnel works as the employee will establish a secure connection through a VPN tunnel; the employee will then login to the VPN client profile using username and password. The authentication is verified with the security policies and the data center. Once the connection is successful, the remote client is connected to the cloud and is ready to utilize the resources and services offered by the application itself. The login of the user will fail if the user is not a valid user, which is checked in the system mainly through the active directory [18]. This type of secure login is highly desirable and is present in the Barracuda application, which was also tested while carrying out this research. It not only protects the user's data, but the users who login into the system through VPN tunnel can be rest assured that they can perform their task without anyone capturing their information.

Encryptions ensure that the data, which are available in the cloud, is secure. Although there are many types of encryption techniques available, attribute-based encryption will provide favorable results. This provides access control with a private key, master key, and ciphers text [19].

Furthermore, as proposed a clear explanation of encryption by Rohit, Rituparna, Nabendu, and Sugata research paper based on security issues in cloud computing, they outline the very important aspect of how the encryption can occur in a cloud-based environment. The argument raised is that that data that is stored in the cloud is secure enough towards any type of security breach. They come up with utilizing homomorphed token, which can help secure data through encrypting by private and public keys, respectively. The trust-based methods for the cloud environment are very valuable towards secure private and public key exchange over a secure seamless synchronized connection. Moving on to further discuss encryption supposedly if data is not encrypted, spoofing attacks can take place. Such attacks can be checked by performing user authentication based on key exchange and even encryption techniques [20]. By enabling encryption sessions with filtering at the entrance of traffic management, such attacks can be avoided. Encryption plays a very important part in securing the cloud services with its unique ability to transform the data into cipher which makes the attackers difficult or almost impossible to alter the data.

Information security relates to gathering the alerts which come about the cloud service monitoring tools. Logs get created for the events. Being a central point, cloud computing is able to handle the information stored and how it gets altered by malicious activity which leads to a crisis situation. If an alert gets ignored, it becomes a golden opportunity for attackers to exploit the cloud services and can access the data of customers. If such a case does happen, the admin must take immediate actions and retrieve data backups. Cloud computing can aid in the seamless transfer of the information to a backup server which will store the information of all the customers. Cloud IaaS is a possible direction of data backup in which data needs to be firmly protected as it should be a specialized cloud-based backup server [21].

Intrusion management looks after the packets coming in and going out of the network. It has got a set of predefined rules which can handle a particular event. A cloud service provider needs to have an intrusion management tool such as anomaly detection. This type of detection system trains itself by observing network behaviors. It identifies the class level for the intrusion whether normal or intrusion, based on the network packets. If an intrusion is found, it should send a warning to the alert or information security system for further action [22]. Hadoop is an open source software, which is becoming popular with cloud administrators. Hadoop is used to distribute processing of big data using MapReduce. MapReduce is a model which can perform analysis very quickly to locate the malicious activity and the area in which the attack occurs [23].

Disaster management in collaboration with disaster recovery relates to cloud data storage in its servers. One must be prepared for it; thus, disaster rescue management can be put in place by the hosting providers in the cloud servers. Attackers can disrupt services by sending malicious requests to the server if there are no strong security policies placed, and they can create downtime of the server as the servers can get overloaded through it. For natural disasters, cloud hosting providers can place their data centers at geographical locations so that if one center gets affected, another will pick up and prevent downtime of services [24].

Looking at an infrastructure point of view, we picked Veeam, a software product developed by Veeam organization itself to replicate, backup, and restore data on virtual machines. It has a lot of capabilities as it pools together one of the leading backup services for a cloud infrastructure. Having the ability to replicate with advanced monitoring, reporting tools, and capacity planning functionality, Veeam is highly desirable to be used for a disaster management tool.

3. Methodology

Based on the research ideas provided, we have used qualitative research method, and the theory we have decided to use is as follows. A local user agent is created by the user to establish a temporary security certificate for safe authentication over a given period of time. This certificate will contain the username, user id, security features, hostname, session times, and other relevant features. Once this is done, the authorization for the user is finalized. As the user will start to use the resources on the cloud, mutual authentication will initiate between the cloud application and user. The application will check if the certificate is valid for the user, a security policy is applied to it. As per the requirements stated by the user, the application will create a list of service resources which will send it to the user. Finally, through an application programming interface (API) security used by the application, the user's session will be fully initiated and connect to cloud services [4, 13].

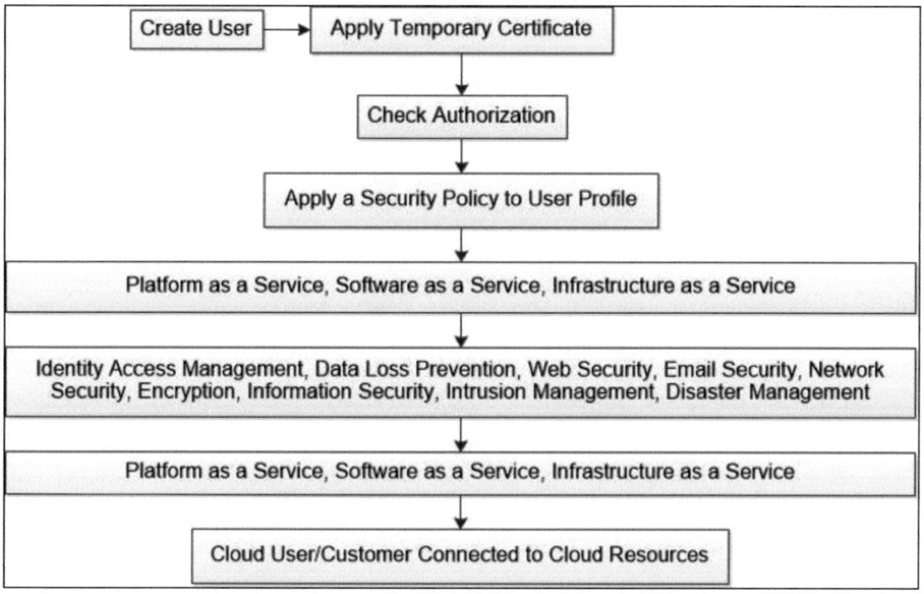

Figure 1.
Model for secure connection with a trusted certificate in a cloud environment.

Figure 1 describes the method for secure connection with a trusted certificate in a cloud environment and describes its successful implementation as well as usage of cloud resources.

Some of the research questions we have identified are as follows:

- Which security protocols you have placed in your cloud architecture to ensure a seamless connection between users does not result in online data theft?

- In case an attack on the cloud service occurs, how will the server mitigate those attacks?

- Is there a disaster recovery management tool in place for the cloud servers? If so then what procedures will be followed to ensure that there is little or no downtime?

- Are the cloud services running behind a trusted firewall? If yes, then how does the firewall report incidents as logs to the administrators and is the firewall artificially intelligent enough to challenge such difficulties?

Our research came up with some cost analysis based on cloud infrastructure. The below details were developed for a cloud-based premise comparing both private and public cloud. Shown below is the cost for Azure sizing based on the requirements; the cost is higher than the private cloud infrastructure with much higher requirements (**Figure 2**).

Shown below is virtual storage area network for a hyper-converged solution which is the most popular infrastructure technology in the current market according to Gartner report. This is very helpful for cloud-based organization to grow as it exceedingly with a lot of resources available for use in the cloud deployment models itself (**Figure 3**).

Based On	Azure Sizing (Public cloud)		Private Infrastructure Requirement
Total vCPU	320		775
Total Memory (GB)	1280		2400
Total Storage (TB)	162.56		200
cost	$	3,974,990.00	1.8 M Production and DR

Figure 2.
Cost analysis with Azure vs. private cloud infrastructure based on resource requirement.

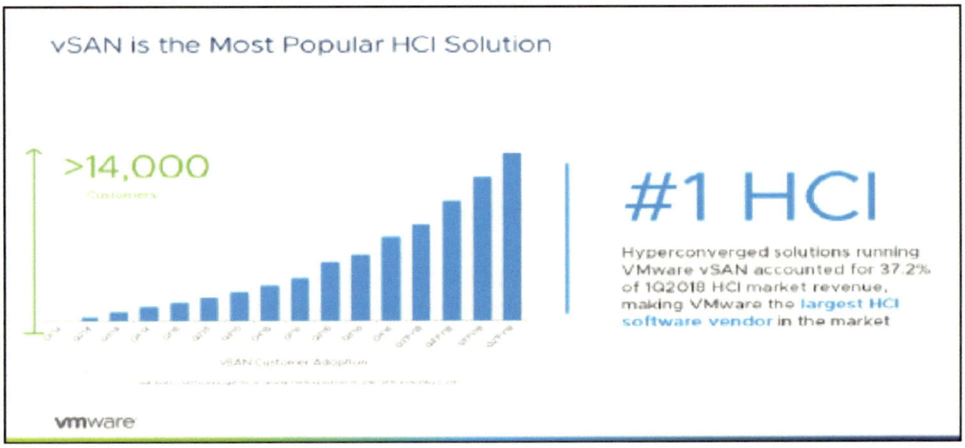

Figure 3.
Virtual storage area network for a hyper-converged solution.

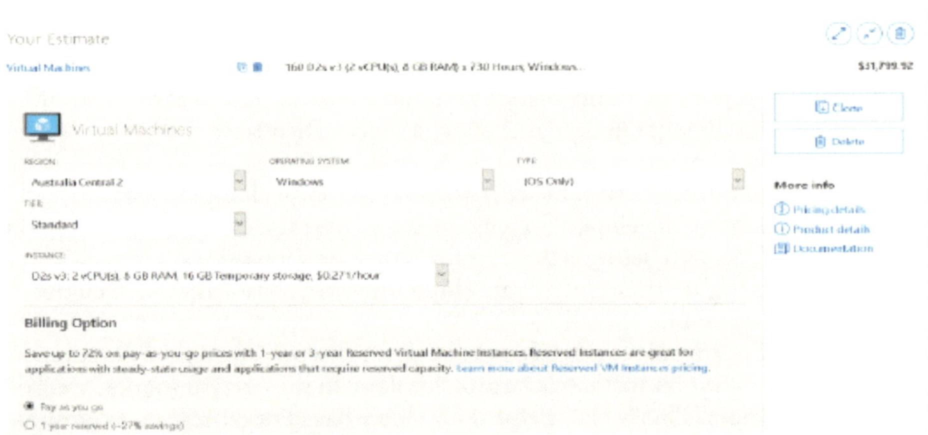

Figure 4.
Billing and purchase interface when requesting a virtual machine in a cloud-based environment.

Name	All Flash FTT1 / RF2
Hardware	Lenovo SR650
Number of Nodes	6
Per Node	Thinksystem SR650
Processor	2 x Intel Gold 6126 12C 2.6GHz
Memory	512GB
Storage Drive	15 x 3.84 TB
Cache Drive	3x400GB + 2x128GB M2
Networking	4 x 10GbE SFP+
Dual Hot Plug Power Supply	Yes
Warranty Hardware	3 Year
VMWare	Included
Cost Per Node	
Cost for Production	$ 900,000.00
Cost for DR	$ 900,000.00

Figure 5.
Costing for a cloud infrastructure with a disaster recovery package.

The screenshot shown below shows the virtual machine on a cloud premise. When a user wants to purchase a virtual machine, the cost related to the resources requested will be shown on the cloud interface and can be upgraded as well when one wants to deploy a virtual machine in their cloud (**Figures 4** and 5).

4. Existing cloud security solutions

The focus of this research is on distributed denial of service (DDoS) attacks on the cloud. The authors researched on existing cloud security solutions and also present an implementable solution focusing on DDoS mitigation for IT infrastructure. The authors define the scope and recommend few focus areas:

- Defending volumetric attacks is a need for cloud components.
- Blocking application-level attacks without submitting SSL Key.
- Deploying acceptable network infrastructure as per IT security policy.

DDoS attack mitigation solutions are discussed here based on design perspective:

a. **On-premise based:** Having a devoted on-premise DDoS attack mitigation answer are first-rate desirable for government entities, financial establishments, and healthcare but not beneficial for all. When the highest stage of safety is mandatory and organizations opt to give as little visibility into their customer facts or approximately their encryption certificate to as few third birthday celebration providers, this could be regarded as a limited scope option. On-premise DDoS devices might store encryption certificates and inspect visitors regionally without any scrubbing, redirection, or inspection. The mitigation device would be required to guard against numerous DDoS vectors like flooding (UDP/ICMP, SYN), SSL based, application layer (HTTP GET/POST), or low and slow attacks. With mitigation structures in house, the proximity to facts center sources is useful, and the systems may be fine-tuned at once by the in-residence IT teams. They have a tendency to have a miles more cognizance to their setup for any adjustments in site visitor flows or from the application servers. Thus, they might have a tendency to have a higher chance of detecting any suspicious traits or visitors requests.

b. **Cloud-based security services:** In providing anti-DDoS and superior mitigation protection in shape of managed security services, many cloud carrier companies offer protection from community floods with the aid of deploying mitigation system on the ISP network edge stage or with scrubbing centers. This involves traffic diversion from the corporation network to detection or scrubbing center. When a DDoS attack starts, human intervention is needed and takes as a minimum of 15–30 minutes all through which the online services are left unprotected and exposed. The cloud-based totally DDoS mitigation service guarantees quantity blocking off of community flood assaults from accomplishing the corporation edge devices or flooding the WAN circuit which is free of volumetric community flood attack. However, there exist glaring problems with a cloud primarily based on DDoS mitigation offerings.

- Cannot discover and block application layer attacks and slow attack.
- Unable to defend stateful infrastructure structures like firewalls or IPS.
- Unable to deal with attacks like software layer attack, state exhaustion, and multi-vector attacks.

c. **Hybrid cloud-based security:** Using hybrid cloud functions gives the best-of-breed mitigation option, where the hybrid infrastructure combines the on-premise in-house setup with DDoS mitigation carriers to act as an included mitigation solution. In hybrid solutions, another option is to use a devoted DDoS mitigation provider's capability in order to detect and block a couple of DDoS vectors. Having public cloud issuer dynamically booms the community pipe bandwidth for the duration of a DDoS attack; takes off a while after being detected, till the time mitigation begins; and saves the on-premise infrastructure from the attack and affecting the provision of its online services. Typical answer is in the course of DDoS attack; the entire site visitors are diverted to a DDoS mitigation issuer's cloud, where it is scanned, scrubbed with the attack visitors getting diagnosed, and removed before being re-routed lower back to the in-residence information middle of the enterprise. Hybrid solutions permits organizations to gain from the following:

- Widest security coverage that can simplest be finished by means of combining on-premise and cloud insurance.
- Shortest reaction time by using an on-premise solution that begins right away and mechanically to mitigate the assault.
- Single touch point during an attack both for on-premise and cloud mitigation.
- Scalability—each tier is impartial of the other and can scale horizontally, in case there is a web application attack spike, adding extra WAF devices to ensure enough WAF capability may be done within the application defense tier without affecting the community tier.
- Performance—on the grounds that requests come in tiers, network utilization is minimized, and load decreased overall.

- Availability—with hybrid solutions, if the first or second tier is down, at least there is one tier left to serve consumer requests. This satisfies the BCP of the organisation.
- Vendor independence—community and application protection infrastructure can setup the usage of hardware structures or even specific software program versions.
- Policy independence—while new policies are implemented at the application defense tier, the opposite tier directs simplest that specific visitors in the direction of the rules until they are established and ready for production use.

5. Proposed DDoS solution

Based on the developing threats and effect of attacks, company firms having their very own cloud services as well as cloud providers put into effect DDoS mitigation using hybrid cloud architecture. When there are multi-vector DDoS attacks targeted at layers 3, 4, and 7, mitigation strategies are essential. These mitigation strategies assist in detecting and preventing volumetric, software, and encrypted assault vectors. By making use of public cloud capabilities to cover for scalability taking on floods and appearing because the first point of defense with community and web application firewalls detecting assault visitors and mitigating the DDoS threats and the SaaS utility, web portals and backend database resides stable in the residence private statistics center. For this research, the experimental environment involved community infrastructure architectures being designed and setup to testing the proposed DDoS solution having the following hardware and software:

- Cisco 4000 ISR Series Routers and Cisco Nexus 5000 Series Switch for routing and switching
- Big IP LTM-4200 for high-performance application traffic load management
- Cisco Firepower FPR-2110, Imperva Web Application Firewall Gateway with Manager Console
- HP DL-360G8 1U-Rackmount with Intel E5, 128 GB DDR3, 32 TB SSD Servers
- VMware NSX-T 3.0 virtualization software on bare-metal HP Server
- SaaS Application running Windows Server 20012 64-bit OS
- Front End Web Portal with .NET supporting 2-Factor authentication
- Back End Database running Microsoft SLQ Datacenter license on Windows 2012 OS
- DDoS Tools for attack simulation: LOIC or Low Orbit Ion Canon, HOIC or High Orbit Ion Canon, Packet Storm (HTTP Unbearable Load King), Are You Dead Yet (R.U.D.Y), Motoma IO's PyLoris, Slowloris and TOR's Hammer

The networks were tested by community and alertness layer attacks with the use of ICMP flooding with a thousand echo requests with increasing buffer size from 3700 to 3805 bytes. The use of DDoS attacks such as LOIC, R.U.D.Y, and slowloris that simulated attacks denied valid users to get admission to the web software portal. When performing the simulated DDoS assaults, the real user monitoring records are taken as the standards, and parameters have been amassed for the logs to assist generated graphs for DDoS attacks. These parameters had been chosen due to the fact that they decide what performance problems the real users are experiencing on the site for the time being in actual time during an assault.

- Average ICMP latency (milliseconds) before and during the course of DDoS attacks on the apps

- Page load response that refers to time the app pages take to load and figuring out where exactly the time is spent from the time a user logs authenticates and logs in to until the page has loaded completely

- App response relates to the percentage time for a page load process to complete

- Status codes are gathered from the HTTP reputation codes the web server makes use of to communicate with the web browser or person agent

6. Performance analysis

6.1 Single-tier network architecture

The first framework was structured and implemented in the form of a single inbound and exit gateway. This mimicked the single-tiered level system including standard system design, directing the interfacing with an online interface containing the front end and back end. This simulated the standard cloud-based condition having a basic standard system configuration actualized in a server farm with hardware devices mentioned in the setup environment above (**Figure 6**).

Using the standard design, single-tiered architecture, multi-vector DDoS attacks were executed as network floods and volumetric and application layer 7 attacks. These critically overloaded and degraded the computing systems leading to access issues for legitimate users. Logs and data gathered for each attack are displayed below for reference (**Figures 7–9**).

6.2 Three-tier network architecture

The second infrastructure was designed as per the proposed design having three unique tiered designs. Each tier has different IP address schemes and communicates with others via site-to-site VPN. This design simulated public and private cloud integration. The first two tiers focused only on security protection against network and application layer attacks. The third tier only focused on access to the hosted SaaS application with database backend:

- The first Tier is built around layers 3 and 4 network defense system for IP and TCP defense using hardware firewalls and load balancer. This tier mitigates ICMP (Ping), UDP, or SYN flood attacks.

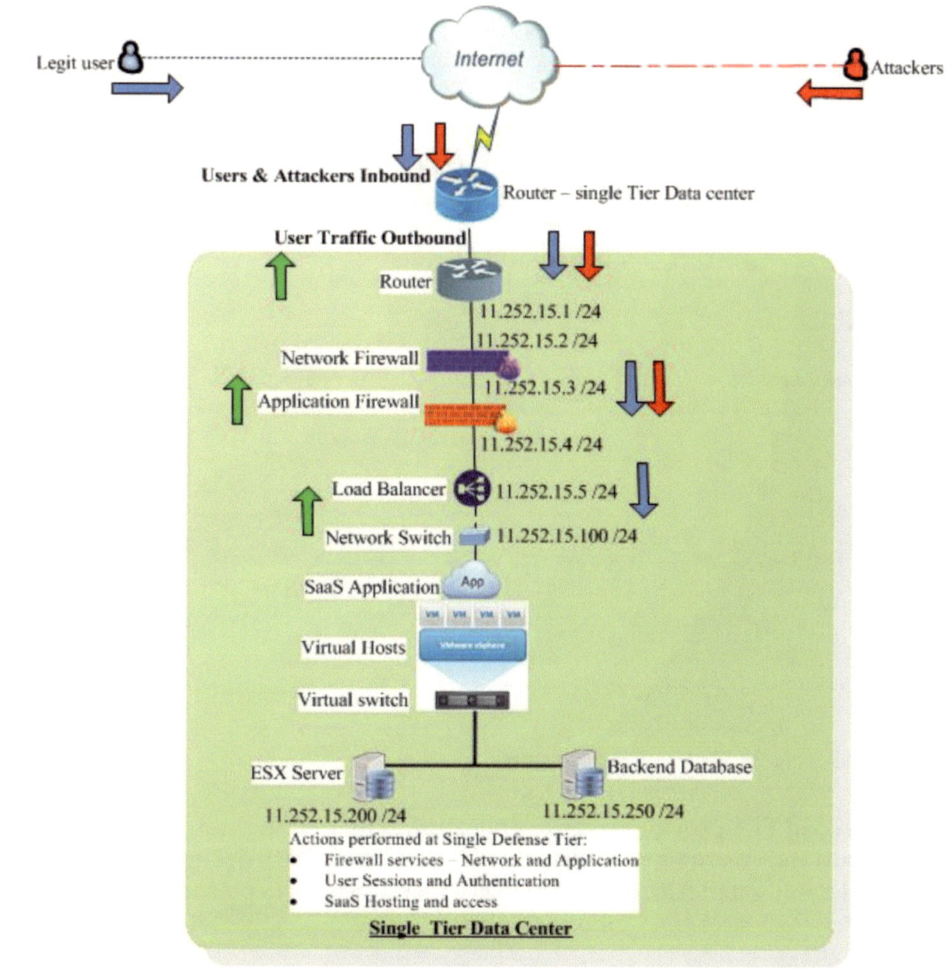

Figure 6.
Traditional architecture design (single-tier).

Before and After Attack Statistics:

Website Response for Network Defense

				Real User Monitoring					
Attack#	Time (pm)	Buffer Size (bytes)	Echo Requests	Average ICMP (ms)	Page Load Response (ms)	Browser Throughput (rpm)	App server response (ms)	Status code	Attack Vector Deta
Attack#1	13:00	3700	1000	6545	45	1800	1636	200	No standard network layer place - single tier archi Ping AppServer -n 1000 Size: 3xxx, Echo request co
	13:30	3750	1000	6670	54	1856	1496	429	
	14:00	3760	1000	6575	55	1727	1624	200	
	14:30	3780	1000	6791	46	1627	1784	200	
	15:00	3790	1000	6583	41	1606	1713	429	
	15:30	3795	1000	6745	55	1806	1686	204	
	16:00	3800	1000	6790	50	1651	1488	429	
	16:30	3820	1000	6794	54	1761	1795	204	
	17:00	3810	1000	6690	47	1800	1833	503	
	17:30	3805	1000	6512	42	1849	1565	503	
	18:00	3820	1000	6692	48	1835	1726	503	
	18:30	3810	1000	6589	50	1635	1570	503	
	19:00	3805	1000	6995	50	1839	1663	503	
Attack#2	13:00	3750	1000	2795	30	1325	1297	200	Network Firewall Defense in Attack vector categories of ICMP/UDP/SYN floods pe
	13:30	3745	1000	2911	32	1327	1243	200	
	14:00	3760	1000	2805	29	1208	1298	200	
	14:30	3780	1000	2963	30	1306	1043	200	
	15:00	3770	1000	2746	29	1235	1097	200	
	15:30	3783	1000	2933	32	1245	1213	200	
	16:00	3780	1000	2988	28	1219	1228	200	
	16:30	3794	1000	2994	29	1270	1064	200	
	17:00	3790	1000	2666	31	1256	1066	200	
	17:30	3789	1000	2934	28	1293	1282	200	

Figure 7.
Single-tier DDoS attack logs.

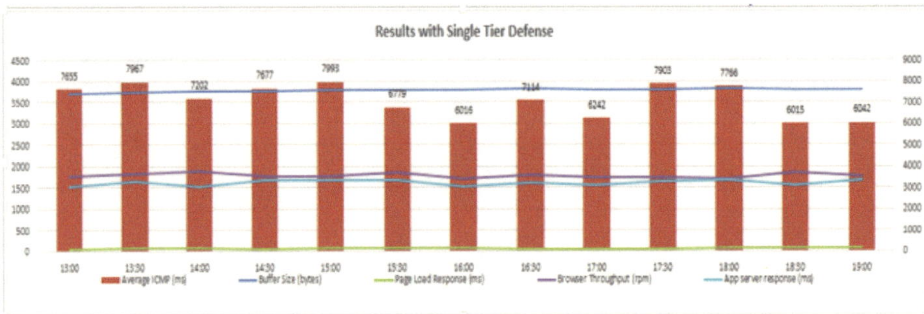

Figure 8.
Single-tier attack results.

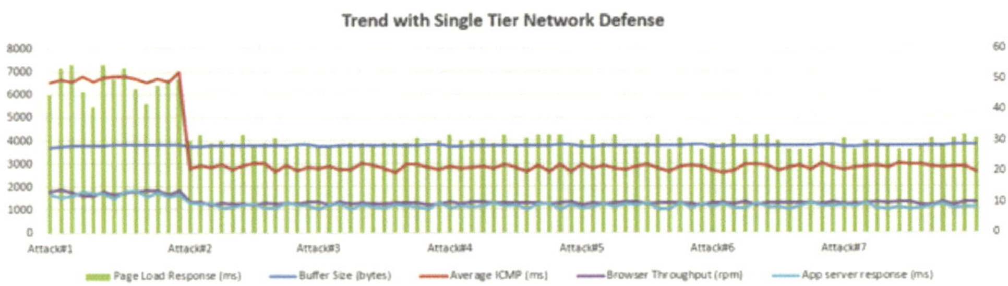

Figure 9.
Single-tier results (with and without network security).

- The second Tier provides layer 7 application defense using web application firewalls and customized load balancing rules along with SSL termination. This tier mitigated ARP spoofing, POST Flood, and DNS poisoning and detected malwares from inbound user traffic.

- Once both network and application attacks are cleaned from the traffic, only legitimate user traffic remained. This traffic is directed to access the private tier cloud (or the third tier), hosting only the SaaS Web portal. After processing and completing the work, user traffic is again reverted to tier 2 for exit instead of tier 1 and following the same traffic route back to the user. This form of asynchronous routing ensures the attackers are not able to execute denial of service attacks that always have the condition of user traffic having the same inbound and outbound gateway and traffic routes (**Figure 10**).

DDoS assaults were performed at first on the single-level system plan and our proposed three-level system structure and assembled result that demonstrate our proposed half-breed cloud configuration having the main level for accepting inbound traffic from clients and aggressors with layers 3 and 4 gadgets and performing system assault alleviation, utilizing a system firewall blocking ICMP floods. The inbound traffic was then permitted to stream to the second level which alleviated application-level assaults utilizing a WAF. Here utilizing F5 and Cisco gadgets intelligently, we had the option to square 80% of the assaults. This was assembled subsequent to contrasting the assault information and single-level system arrangement. The three-level system configuration is executed in a test server farm with Cisco and F5 arrange gadgets for steering, VPN, and exchanging. We utilized VMware and Microsoft operating system servers with a SQL server as backend database to mimic cloud-based SaaS applications. DDoS assault reproductions were performed on the three-level engineering to check the patterns for system and application-level

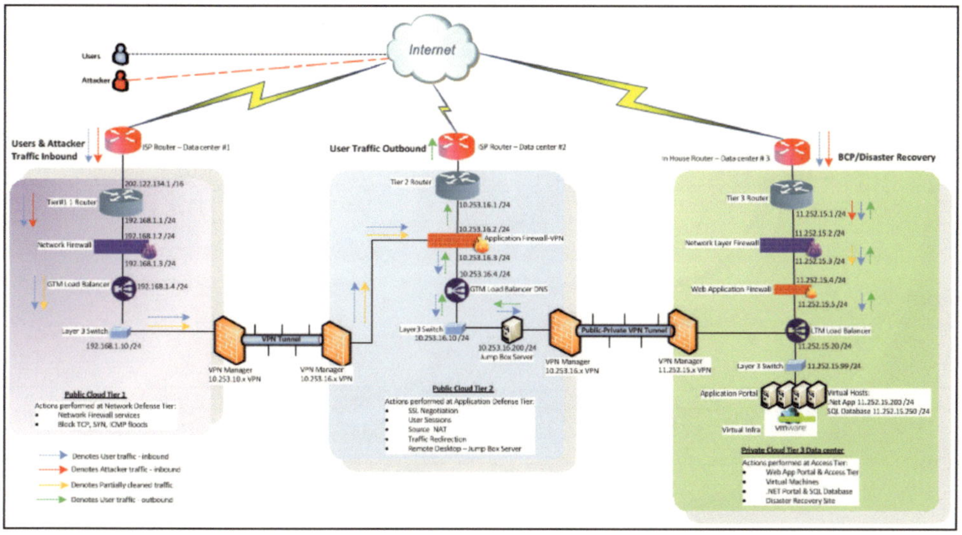

Figure 10.
Proposed three-tier architecture.

Three Tier Attack Statistics

Website Response for Network & Application Defense

Attack#	Time (pm)	Buffer Size (bytes)	Echo Requests	Threads Count	HTTP Flood Get Attack (Wireshark log)	Slowloris socket buildup (perl slowloris.pl)	Average ICMP (ms)	Real User Monitoring: Page Load Response (ms)	Browser Throughput (fpm)	App server response	Status code	Attack Vector Details
Attack#1	13:00	3700	1000	10	GET /app/?id=43779&msg=BOOM%2520HEADSHOT! HTTP/1.1Host: IP	5sec TCP timeout with 1000 socket connections	7655	50	1775	1528	200	No standard network or application layer defense in place three tier architecture Ping AppServer -n 1000 -l 3xxx Size: 3xxx, Echo request count 1000
	13:30	3750	1000	15	GET /app/?id=71309&msg=BOOM%2520HEADSHOT! HTTP/1.1Host: IP	5sec TCP timeout with 1000 socket connections	7967	61	1826	1645	429	
	14:00	3760	1000	20	GET /app/?id=44201&msg=BOOM%2520HEADSHOT! HTTP/1.1Host: IP	5sec TCP timeout with 1000 socket connections	7202	70	1887	1517	200	
	14:30	3780	1000	25	GET /app/?id=62083&msg=BOOM%2520HEADSHOT! HTTP/1.1Host: IP	5sec TCP timeout with 1000 socket connections	7677	58	1773	1683	200	
	15:00	3790	1000	30	GET /app/?id=31881&msg=BOOM%2520HEADSHOT! HTTP/1.1Host: IP	5sec TCP timeout with 1000 socket connections	7993	65	1775	1690	429	
	15:30	3795	1000	35	GET /app/?id=79045&msg=BOOM%2520HEADSHOT! HTTP/1.1Host: IP	5sec TCP timeout with 1000 socket connections	6779	61	1850	1682	204	
	16:00	3800	1000	40	GET /app/?id=74109&msg=BOOM%2520HEADSHOT! HTTP/1.1Host: IP	5sec TCP timeout with 1000 socket connections	6016	63	1704	1534	429	
	16:30	3820	1000	45	GET /app/?id=62360&msg=BOOM%2520HEADSHOT! HTTP/1.1Host: IP	5sec TCP timeout with 1000 socket connections	7114	55	1804	1606	204	
	17:00	3810	1000	50	GET /app/?id=97608&msg=BOOM%2520HEADSHOT! HTTP/1.1Host: IP	5sec TCP timeout with 1000 socket connections	6242	50	1743	1547	503	
	17:30	3805	1000	55	GET /app/?id=91863&msg=BOOM%2520HEADSHOT! HTTP/1.1Host: IP	5sec TCP timeout with 1000 socket connections	7903	52	1751	1651	503	
	18:00	3820	1000	60	GET /app/?id=35656&msg=BOOM%2520HEADSHOT! HTTP/1.1Host: IP	5sec TCP timeout with 1000 socket connections	7766	72	1722	1685	503	
	18:30	3810	1000	65	GET /app/?id=77718&msg=BOOM%2520HEADSHOT! HTTP/1.1Host: IP	5sec TCP timeout with 1000 socket connections	6015	67	1860	1569	503	
	19:00	3805	1000	70	GET /app/?id=24975&msg=BOOM%2520HEADSHOT! HTTP/1.1Host: IP	5sec TCP timeout with 1000 socket connections	6042	64	1772	1674	503	
Attack#2	13:00	3700	1000	10	GET /app/?id=36320&msg=BOOM%2520HEADSHOT! HTTP/1.1Host: IP	5sec TCP timeout with 1000 socket connections	1746	11	1033	776	200	Network & Web ApplicationFirewall Defense implemented: Attack vector categories of attack as ICMP/UDP/SYN floods performed
	13:30	3750	1000	15	GET /app/?id=45075&msg=BOOM%2520HEADSHOT! HTTP/1.1Host: IP	5sec TCP timeout with 1000 socket connections	1574	15	947	859	200	
	14:00	3760	1000	20	GET /app/?id=28843&msg=BOOM%2520HEADSHOT! HTTP/1.1Host: IP	5sec TCP timeout with 1000 socket connections	1548	11	935	850	200	
	14:30	3780	1000	25	GET /app/?id=24543&msg=BOOM%2520HEADSHOT! HTTP/1.1Host: IP	5sec TCP timeout with 1000 socket connections	1798	18	871	715	200	
	15:00	3790	1000	30	GET /app/?id=61274&msg=BOOM%2520HEADSHOT! HTTP/1.1Host: IP	5sec TCP timeout with 1000 socket connections	1795	18	1000	739	200	
	15:30	3795	1000	35	GET /app/?id=47162&msg=BOOM%2520HEADSHOT! HTTP/1.1Host: IP	5sec TCP timeout with 1000 socket connections	1549	15	888	736	200	
	16:00	3800	1000	40	GET /app/?id=67423&msg=BOOM%2520HEADSHOT! HTTP/1.1Host: IP	5sec TCP timeout with 1000 socket connections	1525	10	917	791	200	
	16:30	3820	1000	45	GET /app/?id=41929&msg=BOOM%2520HEADSHOT! HTTP/1.1Host: IP	5sec TCP timeout with 1000 socket connections	1827	12	878	807	200	
	17:00	3810	1000	50	GET /app/?id=65859&msg=BOOM%2520HEADSHOT! HTTP/1.1Host: IP	5sec TCP timeout with 1000 socket connections	1753	18	1029	768	200	
	17:30	3805	1000	55	GET /app/?id=47645&msg=BOOM%2520HEADSHOT! HTTP/1.1Host: IP	5sec TCP timeout with 1000 socket connections	1661	17	908	789	200	
	18:00	3820	1000	60	GET /app/?id=22544&msg=BOOM%2520HEADSHOT! HTTP/1.1Host: IP	5sec TCP timeout with 1000 socket connections	1733	11	1065	892	200	
	18:30	3810	1000	65	GET /app/?id=44380&msg=BOOM%2520HEADSHOT! HTTP/1.1Host: IP	5sec TCP timeout with 1000 socket connections	1635	17	1020	899	200	
	19:00	3805	1000	70	GET /app/?id=36712&msg=BOOM%2520HEADSHOT! HTTP/1.1Host: IP	5sec TCP timeout with 1000 socket connections	1536	11	1093	771	200	
	13:00	3700	1000	10	GET /app/?id=81182&msg=BOOM%2520HEADSHOT! HTTP/1.1Host: IP	5sec TCP timeout with 1000 socket connections	1687	16	906	701	200	
	[illegible]	[illegible]	[illegible]	[illegible]	[illegible]	[illegible]	[illegible]	[illegible]	[illegible]	[illegible]	[illegible]	

Figure 11.
Three-tier logs (network attack).

outcomes after the assaults. ICMP flooding was performed with 1000 reverberation demands each with expanding support size (3600–3800 bytes) with each assault. They made the objective server to react and process the ICMP demands, taking cost of CPU assets and at last square substantial solicitations. Application-level assaults were finished utilizing HTTP Flood GET assault with expanding string check and 1200 reverberation demands utilizing “GET/application/?id=479673msg=BOOM %2520HEADSHOT! HTTP/1.1Host: IP” and moderate attachment development mimicking moderate HTTP assault utilizing perl with logs taken from Wireshark.

Device logs gathered for each attack are illustrated in **Figure 11**.

6.3 Comparing single- and three-tier architectures

DDoS attacks were performed on single-tier and the proposed three-tier infrastructure architecture and results gathered for real user monitoring parameters during the network attacks (**Figure 12**).

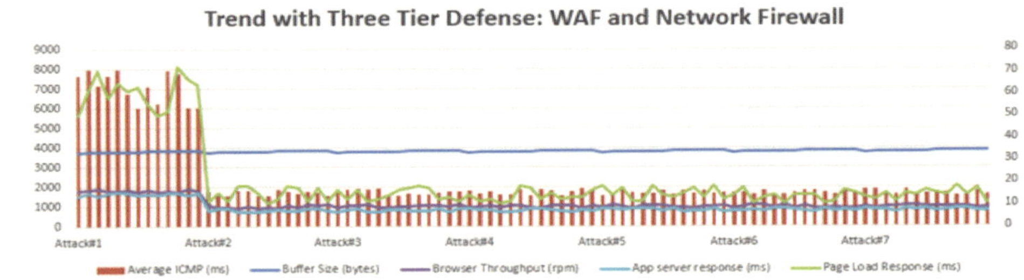

Figure 12.
Network and web defense trends.

					Real User Monitoring					
Attack#	Time (pm)	Buffer Size (bytes)	Echo Requests	Target Server IP	Average ICMP (ms)	Page Load Response (ms)	Browser Throughput (rpm)	App server response (ms)	Status code	Attack Vector Details
Attack#1	13:00	3700	1000	11.252.15.100	6545	45	1800	1636	200	No standard network layer defense in place - single tier architecture Ping AppServer -n 1000 -l 3xxx Size: 3xxx, Echo request count: 1000
	13:30	3750	1000	11.252.15.100	6670	54	1856	1496	429	
	14:00	3760	1000	11.252.15.100	6575	55	1727	1624	200	
	14:30	3780	1000	11.252.15.100	6791	46	1627	1784	200	
	15:00	3790	1000	11.252.15.100	6583	41	1606	1713	429	
	15:30	3795	1000	11.252.15.100	6745	55	1806	1686	204	
	16:00	3800	1000	11.252.15.100	6790	50	1651	1488	429	
	16:30	3820	1000	11.252.15.100	6794	54	1761	1795	204	
	17:00	3810	1000	11.252.15.100	6690	47	1800	1833	503	
	17:30	3805	1000	11.252.15.100	6512	42	1849	1565	503	
	18:00	3820	1000	11.252.15.100	6692	48	1835	1726	503	
	18:30	3810	1000	11.252.15.100	6589	50	1635	1570	503	
	19:00	3805	1000	11.252.15.100	6995	50	1839	1663	503	
Attack#2	13:00	3750	1000	11.252.15.100	2795	30	1325	1297	200	Network Firewall Defense implemented: Attack vector categories of attack as ICMP/UDP/SYN floods
	13:30	3745	1000	11.252.15.100	2911	32	1327	1243	200	
	14:00	3760	1000	11.252.15.100	2805	29	1208	1298	200	
	14:30	3780	1000	11.252.15.100	2963	30	1306	1043	200	
	15:00	3770	1000	11.252.15.100	2746	29	1235	1097	200	
	15:30	3783	1000	11.252.15.100	2933	32	1245	1213	200	
	16:00	3780	1000	11.252.15.100	2988	28	1219	1228	200	

Figure 13.
Single-tier attack parameters.

			Real User Monitoring					
Date	Time (pm)	Threads Count	Average ICMP (ms)	Page Load Response (ms)	Browser Throughput (rpm)	App server response (ms)	Attack detected	ICMP Food Attack
Attack#1	16:00	40	6544	40	1651	1728	GET /HTTP/1.1 404 204	layer defense in place - single tier architecture
	16:30	45	6511	51	1501	1566	GET /HTTP/1.1 404 204	
	17:00	50	6576	37	1555	1728	GET /HTTP/1.1 404 204	
	17:30	55	6525	45	1604	1598	GET /HTTP/1.1 404 204	
	18:00	60	6577	35	1669	1696	GET /HTTP/1.1 404 204	
	18:30	65	6567	38	1594	1575	GET /HTTP/1.1 404 204	
	19:00	70	6402	36	1674	1529	GET /HTTP/1.1 404 204	
Attack#2	13:00	10	4239	24	1132	1053	GET /HTTP/1.1 404 204	WAF Defense implemented: Application layer attack vectors as HTTP attack, Slowloris attack performed
	13:30	15	4113	29	1182	1066	GET /HTTP/1.1 404 204	
	14:00	20	4184	30	1140	1200	GET /HTTP/1.1 404 204	
	14:30	25	4112	20	1219	1000	GET /HTTP/1.1 404 204	
	15:00	30	4233	22	1221	1184	GET /HTTP/1.1 404 204	
	15:30	35	3938	27	1106	1127	GET /HTTP/1.1 404 204	
	16:00	40	4274	25	1258	1012	GET /HTTP/1.1 404 204	
	16:30	45	4269	25	1208	1000	GET /HTTP/1.1 404 204	
	17:00	50	4198	20	1256	1170	GET /HTTP/1.1 404 204	
	17:30	55	4167	26	1204	1176	GET /HTTP/1.1 404 204	
	18:00	60	4318	29	1244	1096	GET /HTTP/1.1 404 204	
	18:30	65	3951	29	1131	1002	GET /HTTP/1.1 404 204	
	19:00	70	3947	27	1203	1022	GET /HTTP/1.1 404 204	
	13:00	10	4059	28	1260	1038	GET /HTTP/1.1 404 204	
	13:30	15	4169	30	1187	1047	GET /HTTP/1.1 404 204	

Figure 14.
Single-tier application attack logs.

6.4 Single-tier logs and data analysis

The below data and graphs illustrate the network firewall and application layer logs and graphs for the DDoS attack performed on single-tier data center architecture in order to determine the resilience for handling DDoS attacks. In **Figure 13** network firewall defense is implemented after attack#2 with ICMP, page load, browser throughput, and application response as the key values.

Figure 14 illustrates real user monitoring values obtained during an application layer attack on single-tier network infrastructure in which application firewall defense is implemented after attack#2 with ICMP, page load, browser throughput, and application response key values.

Results of single-tier architecture attacks obtained before and during the DDoS attack are presented in **Figure 15**. This has the average ICMP, browser throughput, page load response, and application server response.

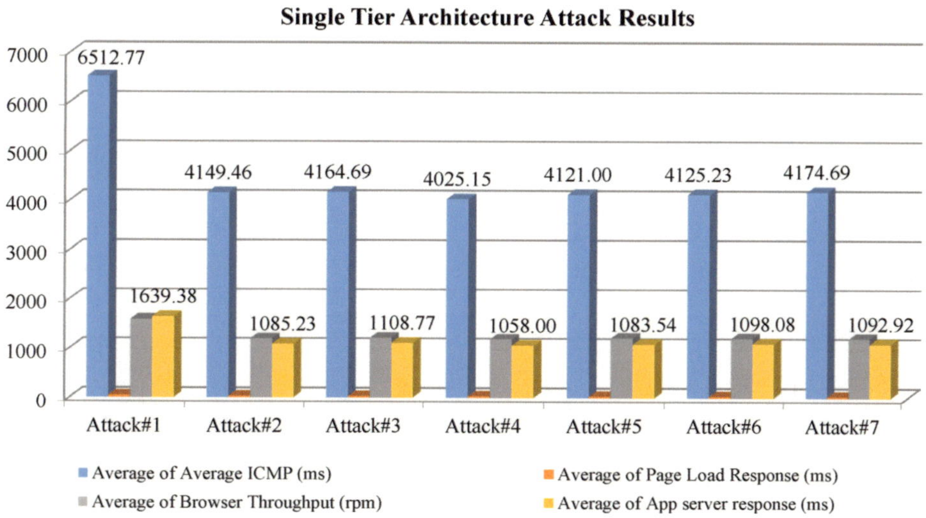

Figure 15.
Single-tier network attack parameters.

					Real User Monitoring					
Attack#	Time (pm)	Buffer Size (bytes)	Echo Requests	Threads Count	Average ICMP (ms)	Page Load Response (ms)	Browser Throughput (rpm)	App server response	Status code	Attack Vector Details
	13:00	3700	1000	10	7655	50	1775	1528	200	
	13:30	3750	1000	15	7967	61	1826	1645	429	
	14:00	3760	1000	20	7202	70	1887	1517	200	
	14:30	3780	1000	25	7677	58	1773	1683	200	
	15:00	3790	1000	30	7993	65	1775	1692	429	
	15:30	3795	1000	35	6779	61	1850	1682	204	
Attack#1	16:00	3800	1000	40	6016	63	1704	1534	429	No standard network or application layer defense in plac three tier architecture Ping AppServer -n 1000 -l 3xxx Size: 3xxx, Echo request count: 1000
	16:30	3820	1000	45	7114	55	1804	1606	204	
	17:00	3810	1000	50	6242	50	1743	1547	503	
	17:30	3805	1000	55	7903	52	1751	1651	503	
	18:00	3820	1000	60	7766	72	1722	1685	503	
	18:30	3810	1000	65	6015	67	1860	1569	503	
	19:00	3805	1000	70	6042	64	1772	1674	503	
	13:00	3700	1000	10	1746	11	1033	776	200	
	13:30	3750	1000	15	1574	15	947	859	200	
	14:00	3760	1000	20	1548	11	935	850	200	
	14:30	3780	1000	25	1798	18	871	715	200	
	15:00	3790	1000	30	1795	18	1000	739	200	
	15:30	3795	1000	35	1549	15	888	736	200	
Attack#2	16:00	3800	1000	40	1525	10	917	791	200	Network & Web ApplicationFirewall Defense implemented: Attack vector categories of attack as ICMP/UDP/SYN floods performed
	16:30	3820	1000	45	1827	12	878	807	200	
	17:00	3810	1000	50	1753	18	1029	768	200	
	17:30	3805	1000	55	1661	17	908	789	200	
	18:00	3820	1000	60	1733	11	1065	892	200	
	18:30	3810	1000	65	1685	17	1020	899	200	
	19:00	3805	1000	70	1536	11	1093	771	200	
	13:00	3700	1000	10	1697	16	906	701	200	
	13:30	3750	1000	15	1867	12	1028	823	200	
	14:00	3760	1000	20	1894	16	1016	857	200	
	14:30	3780	1000	25	1825	11	1093	710	200	

Figure 16.
Three-tier attack logs.

6.5 Three-tier logs and data analysis

DDoS attacks are performed on the designed network architectures and network and application attack results obtained before and after attack scenarios. Network attacks like ICMP flood are done with 1000 ICMP echo requests with each increasing the attack buffer size from 3700 to 3805 bytes. Application attack like HTTP Flood attack is done by increasing the thread count by "*GET / app/?id = 437793 msg = BOOM%2520HEADSHOT! HTTP/1.1 Host: IP*" and slow socket buildup simulating slow web attacks by the use of perl. The logs and Data gathered are gathered from the network firewall; for each attack is displayed in **Figure 16**.

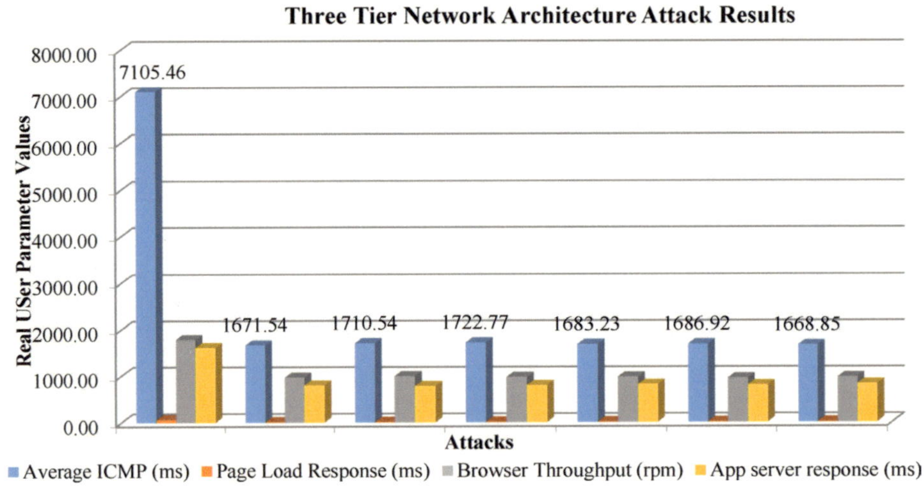

Figure 17.
Three-tier architecture attack parameter results.

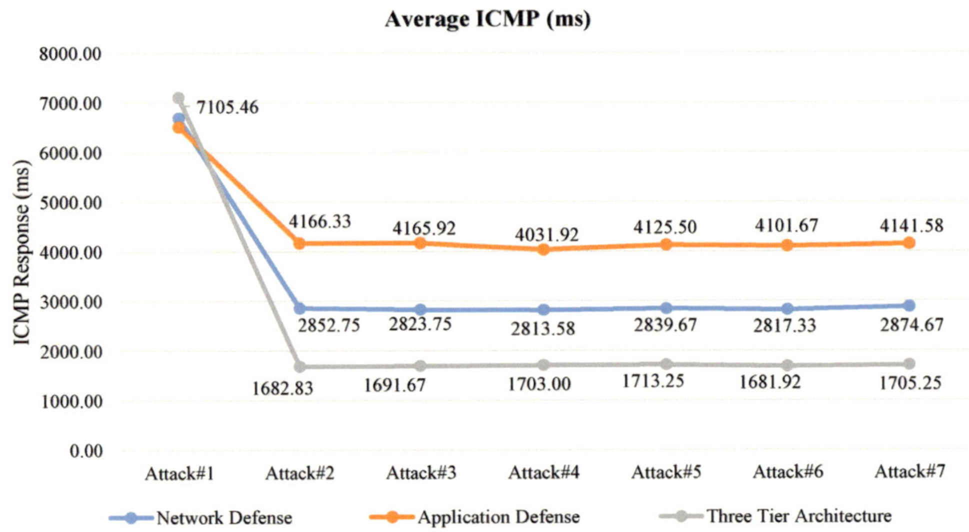

Figure 18.
Real user monitoring for ICMP (single- and three-tier).

Results of three-tier architecture attacks obtained before and during the DDoS attack are presented in **Figure 17**. This has the average ICMP, browser throughput, page load response, and application server response.

The graph in **Figure 18** presents the results of three-tier architecture attacks obtained before and during DDoS attack for ICMP response.

Results of three-tier architecture attacks obtained before and during DDoS attack for page load response is presented in **Figure 19**.

Results of three-tier architecture attacks obtained before and during DDoS attack for browser throughput are presented in **Figure 20**.

Results of three-tier architecture attacks obtained before and during DDoS attack for application server response is presented in **Figure 21**.

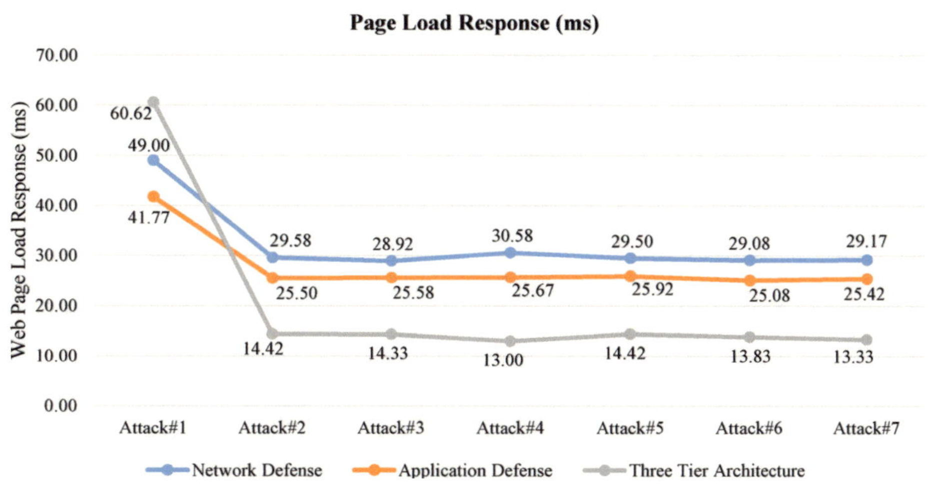

Figure 19.
Real user monitoring for page load response (single- and three-tier).

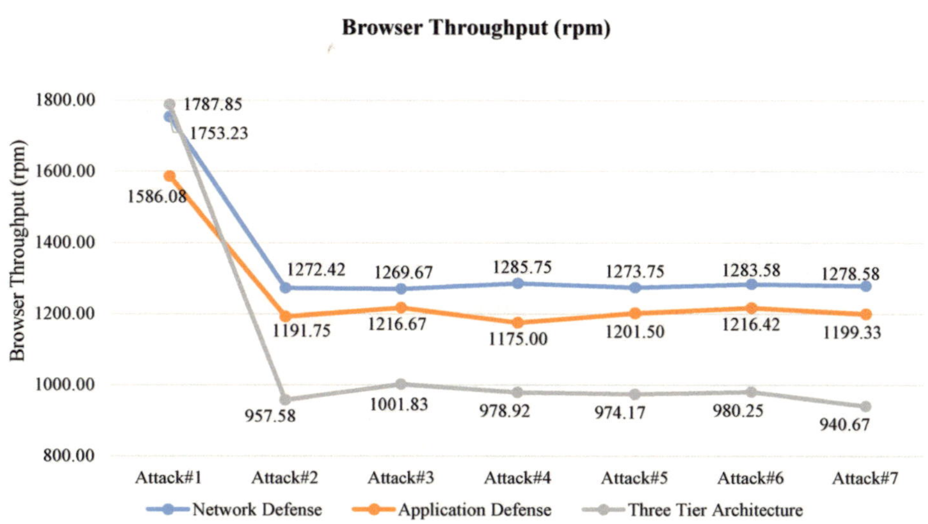

Figure 20.
Real user monitoring for browser throughput (single- and three-tier).

The below graph displays the availability trend metrics obtained after performing the DoS attacks on the two architectures for network and application layer design (**Figure 22**).

6.6 Result analysis

After analyzing the infrastructure, we now focus on what the cloud infrastructure has to offer for implementation (**Figures 23–25**).

1. Firewall → Prevent threats entering the network from outside.

2. Active directory → Authentication, authorization, and group policies for access management.

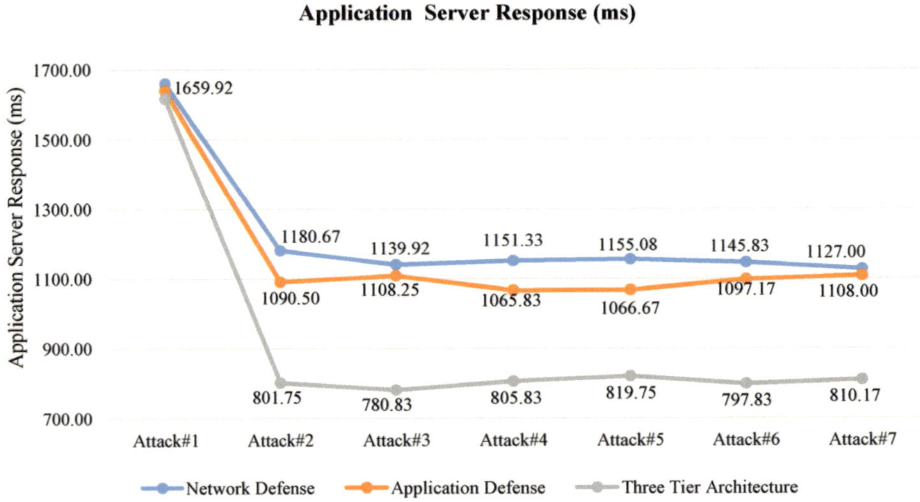

Figure 21.
Real user monitoring for application server responses.

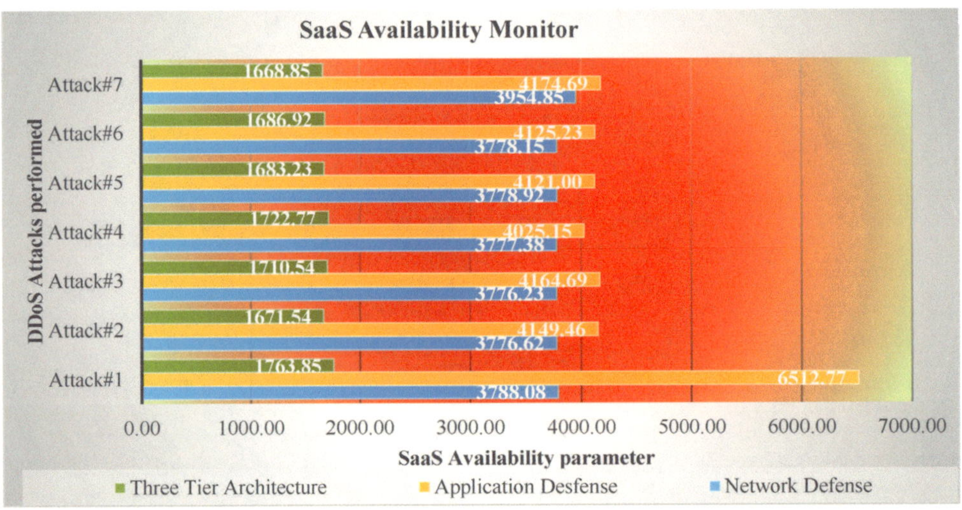

Figure 22.
SaaS availability monitor.

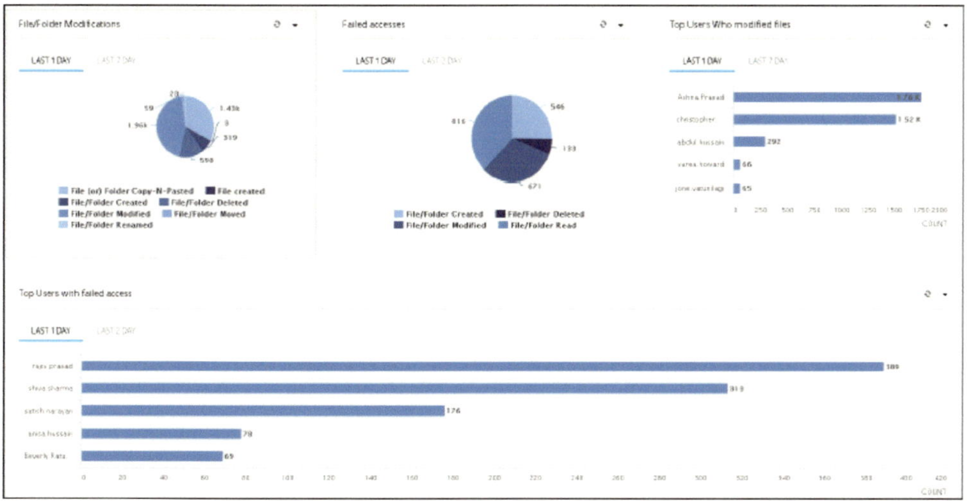

Figure 23.
Populated report for file server access activity on cloud-based premises.

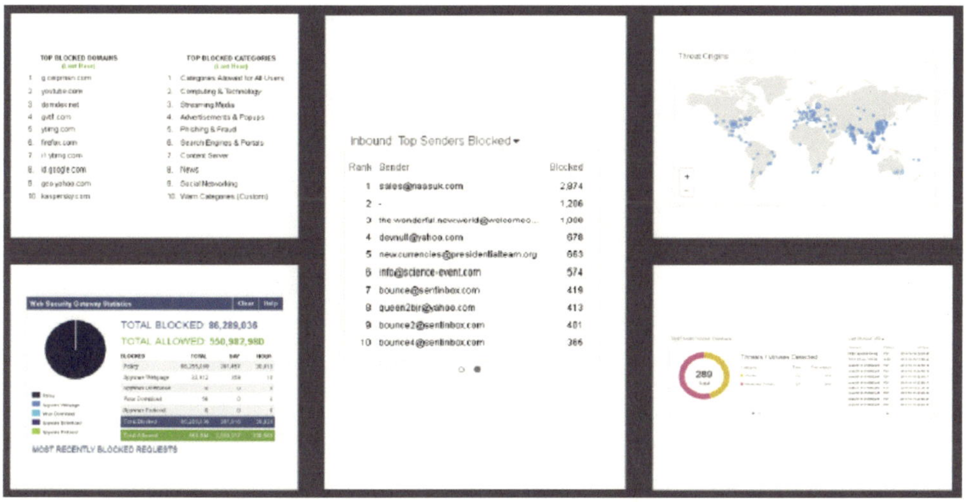

Figure 24.
Threats blocked based on the requests coming into the cloud system.

3. Web application firewall → Protects web servers and manages the incoming and outgoing requests.

4. Web Security Gateway → Web proxy filters manage and monitor websites visited by users in network.

5. Email security → Scans, monitors, and protects emails incoming and outgoing.

Web application firewall (WAF) prevents DDoS attack including SQL injection and XSS attacks to name a few. This is integrated with the implementation as shown in **Figure 26**.

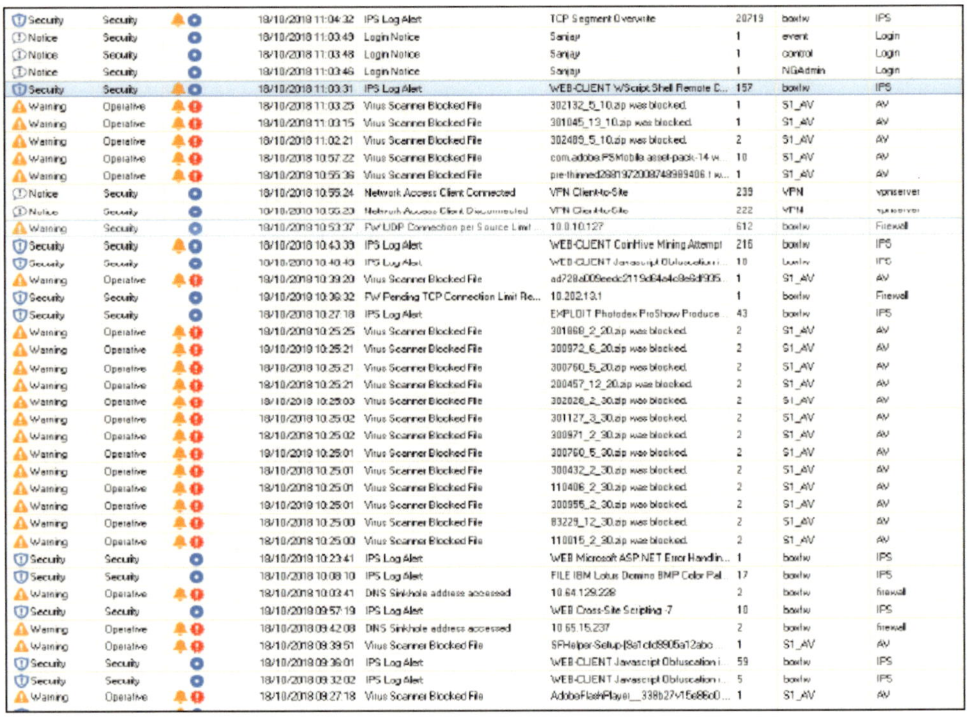

Figure 25.
Logs displaying the threat level with its warning on a cloud-based setup.

Figure 26.
Shows how an attack gets categorized in the WAF.

7. Conclusion

According to our findings and our recommendations having a private cloud, setup is best suited for the larger organizations due to the limitations of going on public cloud infrastructure in terms of bandwidth, data confidentiality, cost of Internet, and cost of its recommended with requirements for infrastructure itself on a public cloud. Setting up a private cloud helps organizations to mitigate risk with confidence and keep 100% control changes to the platform. However, not forgetting the security risks associated with a private cloud infrastructure in areas such as web application firewalls, web security gateways, main gateway firewall, end point security protection, etc., it is essential to have these security appliances implemented with the infrastructure to maintain and protect the cloud environment from outside threats.

Author details

Akashdeep Bhardwaj[1*] and Sam Goundar[2]

1 University of Petroleum and Energy Studies, Dehradun, India

2 The University of the South Pacific, Fiji

*Address all correspondence to: bhrdwh@yahoo.com

References

[1] Rubóczki ES, Rajnai Z. Moving towards cloud security. Interdisciplinary Description of Complex Systems: INDECS. 31 Jan 2015;**13**(1):9-14

[2] Singh SK, Singh DK. Cloud computing: Security issues and challenges. International Journal of Advances in Engineering & Technology. Jun 2017;**10**(3):338

[3] Kajiyama T, Jennex M, Addo T. To cloud or not to cloud: How risks and threats are affecting cloud adoption decisions. Information & Computer Security. 13 Nov 2017

[4] Bunkar RK, Rai PK. Study on security model in cloud computing. International Journal of Advanced Research in Computer Science. 1 July 2017;**8**(7)

[5] Saeed MY, Khan MN. Data protection techniques for building trust in cloud computing. International Journal of Modern Education & Computer Science. August 2015;7(8)

[6] Khajehei K. Role of identity management systems in cloud computing privacy. International Journal of Education and Management Engineering. 2017;**7**(3):25-34

[7] Lazarova V. Managing user access to cloud services by company administrators. TEM Journal. 2016;**5**(3):289-293

[8] Habiba UM. Cloud identity management security issues & solutions: A taxonomy. Complex Adaptive Systems Modeling. 2014:1-37

[9] Shaik K, Narayana Rao TV. Implementation of encryption algorithm for data security in cloud computing. International Journal of Advanced Research in Computer Science. 15 March 2017;**8**(3)

[10] Das N, Sarkar T. Securing cloud from cloud drain. arXiv preprint arXiv:1407.6482. 24 July 2014

[11] Dahiya N, Rani S. Implementing multilevel data security in cloud computing. International Journal of Advanced Research in Computer Science. 1 September 2017;**8**(8)

[12] Xuan S, Yang W, Dong H, Zhang J. Performance evaluation model for application layer firewalls. PLoS One. November 2016;**11**(11):e0167280

[13] Fernandez EB, Monge R, Hashizume K. Building a security reference architecture for cloud systems. Requirements Engineering. 2016;**21**(2):225-249

[14] Alosaimi W, Zak M, Al-Begain K, Alroobaea R, Masud M. Economic denial of sustainability attacks mitigation in the cloud. International Journal of Communication Networks and Information Security. Dec 2017;**9**(3):420-431

[15] He J, Dong M, Ota K, Fan M, Wang G. NetSecCC: A scalable and fault-tolerant architecture for cloud computing security. Peer-to-Peer Networking and Applications. 2016;**9**(1):67-81

[16] Kimbrel JE. Barracuda Launches Suite of Cloud Services for Added Layers of Protection in Office 365 Environments. United States, New York: PR Newswire Association LLC; 2016

[17] Kamla RZ, Yahiya T, Mustafa NB. An implementation of software routing for building a private cloud. International Journal of Computer Network & Information Security. 2018;**10**(3)

[18] Raphiri TV, Dlamini MT, Venter H. Strong authentication: Closing the front door to prevent unauthorised access to

cloud resources. In: ICCWS 2015-The Proceedings of the 10th International Conference on Cyber Warfare and Security: ICCWS2015 Academic Conferences Limited; 24 February 2015. p. 252

[19] Potluri S. Primary methods to address the data security problems in cloud computing. IUP Journal of Computer Sciences. 2016;**10**(1/2):18

[20] Bhadauria R, Chaki R, Chaki N, Sanyal S. Security issues in cloud computing. Acta Technica Corvininesis-Bulletin of Engineering. 2014;**7**(4)

[21] Ramluckan T, van Niekerk B. Security requirements for cloud computing in crisis management. Journal of Information Warfare. 2014;**13**(1):33-46

[22] Al Haddad Z, Hanoune M, Mamouni A. A collaborative network intrusion detection system (C-NIDS) in cloud computing. International Journal of Communication Networks and Information Security. 2016;**8**(3):130

[23] Keegan Nathan S-YJ. A survey of cloud-based network. Human-centric Computing and Information Sciences. 2016

[24] Agarkhed J, Ashalatha R. Security and privacy for data storage service scheme in cloud computing. International Journal of Information Engineering and Electronic Business. 2017;**9**(4):7

Chapter 3

A General Systems Approach to Cloud Computing Security Issues

Svetlana Aristova, Yousef Ibrahim Daradkeh and Petr Korolev

Abstract

An intensive stream of messages about the problem of cloud computing security and a significant number of proposals to mitigate and prevent violation of data privacy and the integrity of the cloud computing environment indicate the relevance and significance of the problem. To bring everything into a certain system is the task of this chapter. We use different methodological approaches in order to find such an integrated solution to the combination of these approaches that, on a unified methodological basis, would allow us to look at the whole range of widening issues of ensuring security and the organization of thinking and activity in the near future. This approach allows us to identify additional problems in this area and outline a program for their development. We try to build a system of methodological design and research over the many private methodologies that authors of articles usually use, relying on the experience of generalizing and concretizing system approaches, and, in particular, expanding geographical and historical boundaries, including system generalizations of intercultural studies and philosophical movements. An attempt is made to disassemble the security problem of cloud computing into a certain number of layers, processes, and technologies of thinking, and to reconnect them into a single whole with the character of thinking and activity.

Keywords: cloud security, general systems methodology, audit

1. Introduction

Many scientific articles, many conferences, many projects are aimed at solving the issue of cloud computing security. Questions suited to this have theoretical and practical significance nevertheless, the problems and significance of this issue have not been identified in its acuteness and clear wording. In all likelihood, the problematization process lacks additional emphasis, namely, the emphasis on determining the positional structure of places for which this issue is significant; emphasis on creating an organizational structure and a system of interactions in which this issue would acquire practical significance and organizational certainty; finally, the emphasis on security and cloud computing as objects with which you can operate and technological chain of operations with objects.

The methodological approach in which we intend to pose a problematization, with the inclusion of the three accentuations described above, was developed for

30 years by a group of developers since 1954. In 1984, it had acquired the form of a pattern of thought activity, such as a scheme of organization of thinking and activity. It has its applications in the context of practical activity of the multidisciplinary group [1, 2]. The practice of applying this methodological principle has become an organizational-activity game. The application of this methodology made it possible to organize extensive research and development material in the field of cloud technology security and to reveal the inadequacy of a number of topics.

The material of our research is 368 articles published in the world press over the past 15 years, with rare exceptions when we turn to earlier works (e.g., on membrane calculations, which were reported in Heidelberg at the 1982 symposium) [3]. It is worth noting that in 2005 in Baltimore, Maryland, the 14th Symposium on Security (USENIX Security 2005) took place, in 2009—the 25th Conference on Computer Security Applications (ACSAC 2009); in the same year in Bangalore, India, an international conference on cloud computing took place. With a relatively stable number of conferences devoted to this topic, it is worth noting the surge in interest in this topic in 2012, 2014, and 2016. The topic of cloud computing has been especially updated since 2016. John Willey & Sons published the Computer Computing Encyclopedia (2016), held the 9th International Conference on Utility and Cloud Computing ACM, 2016; Honorulu hosted the 10th International Conference on Cloud Computing (CLOUD), IEEE, 2017; Workshop on Cloud Computing Security took place in Dallas. Among other things, 31 conferences in Taipei, Taiwan 2017, and the IEEE International Conference on Cloud Engineering (IC2E 2018) were held; the book edited by W Rivera "Sustainable Cloud and Energy Services: Principles and Practice" (Springer International Publishing, Cham, 2018) has wide expansion. The emphasis of research and development is moving toward the development of the computing industry, its applied aspects, such as advanced computing and IT, convergent cognitive IT, Security and privacy (SP), parallel and distributed processing, offensive technologies, Internet of Everything, defined network and network function virtualization, moving target defense, Internet of Things (IoT), and dependable computing (15th European Dependable Computing Conference 2019).

2. Literature review

The topic of cloud computing security has a wealth of development and generalization material. Farnga [4] provides a risk assessment table for the cloud computing environment, introducing three attributes: Probability of Vulnerability (improbable 1, probable 2, occasional 3, frequent 4); Risk Impact (negligible A, marginal B, critical C, fatal D); and Severity Category (low 1A, 2A, 1B; medium 3A, 4A, 2B, 3B, 1C, 2C; high 4B, 3C, 4C, 2D, 3D, 4D). He marks vulnerabilities (Session Riding and Hijacking 4D, Virtual Machine Escape 2D, Reliability and Availability of Service 2C, Insecure Cryptography 3C, Vendor Lock-in, Data Protection and Portability 2C, Internet Dependency 3A) and prescribes protocols to prevent them (**Table 1**). He also defines threats and marks them: Abuse and Nefarious Use of Cloud services 4A, Insecure Interfaces and APIs 3C, Insider threat 3D, Data Loss and Leakage 2D, Account or Service Hijacking 4B, Unknown Risk profile 3D, and recommends risk mitigation protocols. Operational risks (4D) are the following: implementing too quickly, integration issues, moving the wrong data or applications to the cloud, compliance, and cost implications.

In addition to such purely practical manuals, literature is replete with a variety of areas of research and development in the field of cloud computing. Here are some of them. Wazid et al. [5] view fog computing as an add-on for cloud computing,

Protocol	Description
AC-2	Account management
AC-5	Separation of duties
AC-6	Least privilege
AC-10	Concurrent session control
AC-11	Session lock
AT-2	Security awareness
SC-13	Cryptographic protection
SC-23	Session authenticity
SC-24	Fail in known state
SC-27	Operating system—independent applications
SC-28	Protection of information at rest
SI-3	Malicious code protection
SI-4	Information system monitoring
SI-7	Software, firmware, and information integrity
SI-13	Predictable failure prevention
SI-14	Nonpersistence
SI-15	Development process, standards and tools
CM-2	Baseline configuration
CM-6	Configuration setting
CM-7	Least functionality
CA-7	Continuous monitoring
CA-8	Penetration testing
CP-11	Alternate communications protocols
PM-12	Insider threat program
AR-4	Privacy monitoring and auditing
DM-2	Data retention and disposal
AU-12	Audit generation

Source: Farnga [4].

Table 1.
Risk mitigation tools.

which is why fog computing inherits all of the security and privacy issues inherent in cloud computing. They report that they have developed a new key management and user authentication security scheme, named by them as SAKA-FC. The development is based on the well-known Real-Or-Random (ROR) model and the Automated Validation of Internet Security Protocols and Applications (AVISPA) tool. The scheme finds its effectiveness for its use in smart devices with a one-way cryptographic hash function. Guan et al. [6] discuss issues related to data security protection of personal data in fog computing. Fog computing, as an intermediary layer between the cloud and the end user, according to the authors, is precisely the solution to the problems of cloud computing security. This chapter discusses the design of a solution to ensure data security and privacy in fog computing. It is reported that simply transferring the protection techniques used in the cloud to the fog does not produce the desired effect. Alamer et al. [7] explore the safety of road traffic systems (CVCC) by modeling a network of cloud-based moving mechanisms in the form of a two-phase heterogeneous public good game (HPGG Model). This development helps develop security solutions for communications such as vehicle-to-vehicle and vehicle-to-infrastructure, as well as the ability to integrate smart devices and various CVCC applications. Sharma et al. [8] considers that the best solution to protect the cloud from attacks is to use intrusion detection systems (IDS) in combination with different detection techniques. The chapter presents various architectures based on the cloud IDS, which are embedded cloud environments to address various security issues. Fadi and Hemayed [9] provide a literature review of the proven clouds that are used in infrastructure as a service contracts. The authors argue that the integration of the new technology, which is trust computing, with cloud computing can be provided by the proposed architectural solutions of the infrastructure as a service and on the grounds on which user trust in cloud service providers arises. Remote certification and a trusted virtual domain are important security considerations for cloud computing. A security model based on the separation of the security domain was proposed by Xu and Zheng [10] for telecommunication services. Security measures cover the storage, processing, and transmission of data in the cloud. Instead of traditional computational models of cryptographic protection, Maharajan and Paramasivan [11] offer molecular protocol (DNA) membrane computing protocol. Qui and Kung [12] as invited editors provide a clear overview of 14 articles on the topic of cloud computing security. They were selected from 57 proposed articles. The urgent need for the development of techniques and tools for cybersecurity of clouds is noted. Among the authors of the articles are noted groups Ali, Zhang, Lee, Li; Fowley, Chen, Islam, and Sha; Chi, Luna, Awad, Cafaro, Zhang, and Xu are well known in the professional community. The various cybersecurity techniques and tools described by these research and development teams are described. Xu et al. [13] analyze the relationship between openness and cloud security by addressing the results of this analysis (quantitative methods and qualitative analysis of investments in security and openness) to cloud computing providers to adopt an optimal investment strategy for openness and security. Sajai et al. [14] offer a hybrid technology of cryptographic data protection in the cloud, combining homographic and blowfish algorithms. Wei et al. [15] noted that, according to data released by the Cloud Security Alliance (CSA) and the Institute of Electrical and Electronics Engineers (IEEE), there has been an increasing involvement of cloud computing for manufacturing purposes. The authors draw attention to the complex nature of the cloud system, introduce indicators for evaluating the cloud computing system, and propose a rule of believe (BRB)-based model for predicting the safe state of the cloud. This model combines a system of expert assessments and long-term data analysis and has three levels focused on the safety of equipment, software, and services. Bhandari and Zheng [16] describe 12 cloud security threats, such as data breach, insufficient identity, credential and access management, insecure interface and APIs, system vulnerabilities, account hijacking, malicious insiders, advance persistent threats, data loss, insufficient due diligence , abuse and nefarious use of cloud services, denial of services, and shared technologies issues. Donno et al. [17] analyze the situation in which every "thing" is connected to the Internet. From the point of view of security, the technological revolution brings with it many dramatic moments. The authors offer a comprehensive overview of cloud computing security issues in the Internet of Things era. The bibliography for the article has 149 sources. Matheus and Vieira [18] at the student forum of the 15th European Dependable Computing Conference (EDCC 2019) presented a four-step sequential change model for a cloud architecture model, extending the availability and security model to a holistic cloud presentation model and security assessment using Moving Target Defense.

This diversity is striking in its diversity and, in order to deal with the fundamental, essential side of the problem, the proposed methodologies are of little use due to their inconsistency. But the first layer of ideas is nevertheless lined with them. So, we have a certain field of practice and a subject built on it, which combines problems and tasks, knowledge, models and experiments, languages and methods. For the purpose of generalizing and translating this design into a megamachine's plan, it is worth building a block of private methodologies, as well as blocks of methodological design, research, and auto-reflection. So, in relation to the world of things covered by the new digital context, the following can be said. (1) The Internet of things, this new era in the sociocultural development of mankind, requires a certain environment in which each thing has its digital counterpart. A new layer of material organization is taking shape when, by referring a digital double to a thing, the latter reveals itself not only in the localities, but also in new qualities, in new directions of its use. (2) This environment, being distributed everywhere, resembles a certain smart layer covering the entire terrestrial space of things, it contains the systemic representation of a thing in its dynamics, the totality of all kinds of actions with a thing. (3) Speaking of the world of things, we include their interacting and developing aggregates in it, we expand the world of things to the world of activity, with the help of which things are not only created and consumed, but also undergo the influence of constructive thought. In this sense, we can talk about the world of thought activity. Ideal objects of scientific substantive thinking, cult rituals, customs of communication and polemics—all these—form this intelligent world of activity. (4) In a sense, the Internet of things with its infrastructure and cloud computing platforms should be considered one of the forms of such a world of thought activity. (5) An industrial structure is taking shape in which a thing is made with its digital counterpart. This makes the thing more convenient and at the first stage more expensive. Issues of owning a thing, transferring it by inheritance, its commercial use, that is, giving a thing a certain active beginning, can also have their object form and their digital counterpart. (6) Customs, ethics of relationships, trust, and control are things in our world. How will they evolve with the development of the digital era? What customs need rethinking? Is it always necessary to duplicate the predominantly conflict-free world of things in the world of cloud computing? Is activity based on principles other than the order of the real world? Data in our world, everywhere is gaining special significance, both in business and in the social environment. He pointed out that only 17% of companies make data-driven decisions. By 2025, the global data volume will grow 10 times and reach 163 Zettabytes (one Zettabyte contains 10 to the 21st power of bytes), and most of these data will be generated by enterprises, not consumers. Sixty percent of the world's data will be created by business organizations. Almost 20% of all data in the global infosphere will play a critical role in everyday life, and about 10% will be "supercritical." Almost 90% of all data will require a certain level of security, but only half of them will be really protected. The growth of big data and metadata will lead to the fact that by 2025 each average inhabitant of the Earth will begin to interact with devices connected to networks about 4800 times a day, according to one interaction procedure every 18 seconds. The share of the global information sphere under analysis will increase by 50 times compared to the current one, reaching 5.2 ZB; and the amount of data analyzed with the participation of cognitive systems will grow 100 times, amounting to 1.4 ZB. Almost 20% of the data generated will be real-time information, with more than 95% of the data coming from IoT devices [19]. These estimations mean that the problem of security of calculating and computing media will remain actual one.

3. Methodologies

The data used in this study is taken from open sources. The methodologies used by researchers can be expanded to private system-structural methodologies of management, sciences, engineering, and production. We are trying to look at the situation associated with the use of cloud computing from a wider angle by introducing another add-on—the general methodological system-structural design and prospecting ([1], p. 103). You can implement several plans: (1) look through all the literature and write an attitude toward it, making some kind of system generalizations and arrangements; (2) write independently of the literature your understanding of the situation and construct a certain field for assembling sources and identifying niches for their subsequent filling; (3) and design the futures.

When we retrospect to the past studies on this issue, we focus on the following passage from G. Schedrovitsky's paper of 1981, titled "Principles and general plan of the methodological organization of the systems and structures studies and elaborations" ([1], pp. 88-114). He wrote: We distinguish eight projects in which the system principle is developing.

This is a project for the development of specific sciences and areas of engineering and practice due to systemic representations, concepts, and methods of analysis [20–22]; three projects of the "general theory of systems," similar to the natural science theories, such as physics, chemistry, biology [23–31], similar to traditional mathematics such as geometry and algebra or Shannon information theory [22, 32–37], according to the type of metamathematics in the sense of Hilbert and Klinn [38, 39], a practical methodology or methodology of the type of disciplines such as the study of operations, decision analysis [40–42], an engineering methodology such as systems engineering of Good and Mackoll [43–47], the so-called system philosophy [48] and system-structural methodology as a division of the general methodology [49–58].

The first seven proposals have a historical prototype already implemented on another material. This is their forte. At the same time, in our opinion, this raises major objections. When each of the participants in the systemic movement offers his own professional solution to systemic problems, he acts as an agent of the already existing and functioning sphere of thinking and activity—science, engineering, mathematics, philosophy, etc. He has formed as a "system engineer" inside of the sphere, and by virtue of this, he is always connected and limited to that particular cultural and historical situation in which he understood the meaning and importance of systemic problems and tasks. Consequently, in the final analysis, he always only develops, due to systemic means and methods, the professional organization of his initial thought activity. However, it is well known (and can even be considered universally recognized) that the systemic movement has developed and is developing as an interdisciplinary and interprofessional formation. This means that it must form and create an organization that goes beyond the scope of each individual scientific discipline and each individual profession. Consequently, the system movement in its formation and development should take into account the contemporary sociocultural situation as a whole, and proceed from an extremely wide understanding of the possibilities and prospects of its development.

In our opinion, in the current sociocultural situation, at least eight points that have the most direct connection with the systemic movement can be distinguished.

The first of these is the process of an ever-deepening differentiation of sciences and professions. Progressive in the eighteenth and nineteenth centuries, it has now led to the design of a mass of isolated sciences, *S* and *PM* (see **Figure 1**), each of which develops almost independently of the others. These subjects now not only organize but also limit the thinking of researchers. Receptions and ways of thinking,

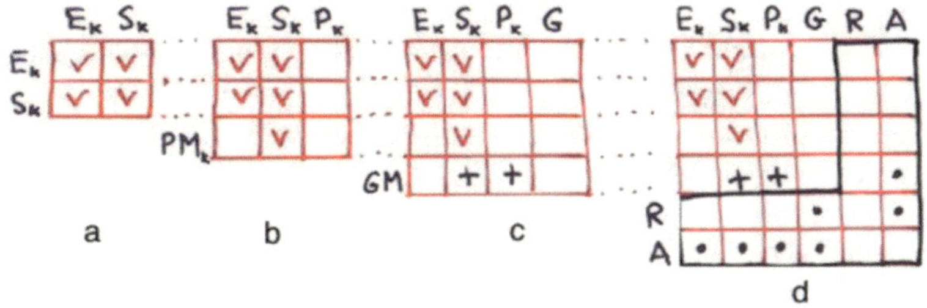

Figure 1.
Four squares, diagonal (a), triagonal (b), quadrogonal (c), and hexagonal (d) images depict a form of research organization of a certain set of practices E_K.

new techniques and new methods created in one subject do not apply to others. Each science creates its own ontological picture, which is not compatible with the ontological pictures of other objects. All attempts to build a unified or at least connected picture of our reality run up against great difficulties.

The second point is the existence of highly specialized transferring channels of fragmented culture. The mathematician does not know and understand physics well, not to mention biology or history. The philologist, as a rule, does not know mathematics and physics, but is equally poorly versed in history and its methods. Already at school, we begin to divide children into those who are capable of mathematics and capable of literature. The idea of general education is increasingly being destroyed by the idea of specialized schools.

The third point is the crisis of classical non-Marxist philosophy, caused by the realization of the fact that this philosophy has lost its means of controlling science and has lost the role of coordinator in the development of sciences, the role of mediator, transferring methods and means from one science to another. This circumstance became clear already in the first quarter of the nineteenth century and became the subject of special discussion. K. Marx and F. Engels paid much attention to it in their works, which redefined the functions of philosophy in relation to the natural and human sciences. The loss of a direct connection with philosophy led various sciences to develop their own forms of awareness, their own individual philosophy. This has provided the basis for various forms of positivism, and in recent times has given rise to the so-called "scientism."

The fourth point is the design of engineering as a special activity that combines design with various forms of quasi-scientific analysis. The traditional academic sciences, which were developed in many ways immanently, were divorced from new areas of engineering, and this forced engineers to create new types of knowledge systems that did not meet traditional patterns and standards. Information theory and cybernetics are just the most striking examples of such systems. At the same time, the problem of the relationship between design and research appeared and began to be intensively discussed.

The fifth (very important) moment is the continued isolation within the activity and the isolation of various production technologies, which acquire self-sufficient importance and become, as it were, a new principle and an objective law in the organization of our entire life activity and ultimately subordinate to ourselves both the activity, nature and behavior of people. Maintenance of these technologies is becoming the primary need and almost the main goal of all social activities. At the same time, technological forms of organizing activities are constantly formalizing and becoming increasingly important, which apply to thinking.

The sixth point is the formation, design, and partial isolation of design as a special kind of activity. As a result, the issue of the relationship and correlation of the actual design and research developments arose even sharper. Designing directly and with all acuteness ran into the problem of the ratio of natural and artificial in the objects of our activity [45, 51]. None of these problems has been resolved within the framework of traditional sciences.

The seventh point is an increase in the importance and role of organizational and managerial activity in our entire social life. Its effectiveness depends primarily on scientific support. However, traditional sciences do not provide the knowledge necessary for this activity; this is primarily due to the complex, synthetic, or, as they say, complex, the nature of this activity and the analytical, or "abstract," nature of traditional scientific disciplines.

The eighth point (also especially important) is the appearance of a new type of science, which could roughly be called "complex sciences." These include the sciences serving pedagogy, design, military affairs, management, etc. Now these complex types of practices are served by chaotic agglomerations of knowledge from various scientific disciplines. But the complexity and versatility of this practice, its orientation at the same time both on normative, artificial, and on implementation, natural plans of activity require a theoretical unification and theoretical systematization of artificial and natural knowledge, which cannot be achieved.

Contemporary situation in general systems theory looks like the same described by G. Schedrovitsky in 1981 ([1], pp. 88-114). Some additions to this domain make it more clear. An article in Wikipedia [59] pays attention to the point that systems theory is the interdisciplinary study of systems. "The goals of systems theory are to model a system's dynamics, constrains, conditions, and to elucidate principles (such as purpose, measure, methods, tools) that can be discerned and applied to other systems at every level of nesting, and in wide range of fields for achieving optimized equifinality."

Dubrovsky ([60], p. 20) makes endeavor to reinterpret the system approach of G. Schedrovitsky. Zilberman [61] identifies six types of cultural traditions. The Vedanta scheme characterizes the Indian type of tradition (methodological thinking as actually "understanding"), the mimansa scheme is the Tibetan type (conceptual or "substantive" thinking), and the Vaisheshika scheme is the new European type "imaginative," axiological, or historical thinking. Further, the nyaya scheme characterizes the Hellenic type of tradition (organizational, axiomatic, mathematical-theoretical, formal-logical thinking), the Sankhya scheme—the Chinese type ("projective," "preformative," praxeological thinking), the yoga scheme—the Japanese type (phenomenological, or existential thinking). All these complex calculations, however, are necessary for Zilberman to label or draw another universal picture of world cultures and civilizations, in the manner of Spengler or Toynbee. Here, rather, a method of intercultural interaction is proposed, with the help of which one can describe any system of culture and at the same time not fall into naturocentrism. By modifying the types of philosophical systems, Zilberman focuses on the ideal of complete modalization of all philosophies so that a "sum of philosophy" arises and the true history of this discipline begins. The thread of modal methodology lies in the fact that for the first time it consciously and intentionally refers not to versions of "reality" as unconditionally natural and therefore problematic for consciousness, but to typological thoughts that it improves. In this sense, the modal methodology plays the role of Philosophia Universalis [61].

4. Results and discussion

4.1 Preliminary data

From our point of view, the specific organizations that solve these problems are the organizations of methodological thinking and methodological work, which should not be identified either with the philosophical proper or with the scientific forms of organization of thinking and activity.

The methodology takes into account the differences and the multiplicity of different positions of the figure in relation to the object; hence, work with different ideas about the same object, including different professional ideas, in this case, knowledge itself and the fact of their multiplicity, are considered as an objective moment in the research situation.

Figure 1 depicts four squares, we will call them diagonal (a), triagonal (b), quadrogonal (c), and hexagonal (d) images; they depict a form of research organization of a certain set of practices E_k. By practices, we mean the entire existing set of activities related to the use of cloud computing, as well as ensuring the security of the use of the cloud. These practices are described within the framework of the S_k description languages that cover them. Note that these languages are different, and translation from one language to another is hardly possible. Means and methods, as well as a description of problem areas and their resolution tasks, are provided by a layer of partial applied methodologies (in the figure they are designated as PM_k). The triagonal image (b) defines the organizational form of the structure of the simplest scientific subject.

In special logical and methodological studies (see, in particular, [62]; pp. 106-190), it was established that in every scientific subject there are at least nine different epistemological units: (1) problems, (2) tasks, (3) "observable facts," (4) "experimental data," (5) the totality of the general knowledge that is built in this scientific subject, (6) ontological schemes and pictures, (7) models, (8) tools (languages, concepts, categories), and (9) methods and techniques. This is a set of basic blocks of a scientific subject.

Our task is to find a solution to the problem of ensuring the security of cloud computing in some unified system language. To this end, we turn to the quadrogonal image, introducing another layer—the general system-structural methodology (in the figure, it is indicated by the letters GM). As part of this add-in, work is underway to design and prospect the system area including as a part PM_k, S_k, and E_k. To the extent that the diagonal image is not complete, the same tetragonal image is also not complete. Let us explain how this layer is built. Following the "Principles and basic schemes of organizing systemic structural studies" ([1], pp. 88-114), we turn to the hexagonal image. It adds two more add-ons, which we marked with the letters R and A, methodological reflection, or auto-reflection (metamethodological area) and audit (the type of methodological research by which the layers of practice, descriptions, applied and general methodology are added and adjusted). The problem areas identified in the layer of private methodologies are also accompanied by a general description that includes, in addition to the technical, engineering, and managerial contexts (determined by the practitioners of experiences), a certain general sociocultural context. This is generated by audits at all levels of the methodological organization, from specific practices to the organization of the design and futures of partial methodologies.

We used the kinematic scheme [60] for organizing methodological work in the field of cloud computing security. The kinematics of the scheme lies in the fact that it combines several methodological schemes, both early in appearance and subsequent ones. The scheme by which David Zilberman tries to build a modal methodology as a sum of methodologies (1973) is supplemented by a scheme of thought activity (1980)([1], pp. 281-298), a scheme of organizing a system-structural methodology (1981) ([1], pp. 88-114), and scheme (2016) that we use when working on the theme of Observation and Audit of the Processes in Experiences with Uncertainty [63] and the scheme (2000) when we were working on the topic of Reflexive Control [64]. We also used our ideas about the inclusion of thinking technologies, such as problematization, objectification, self-determination, and schematization, in this kinematic scheme, which has an enneadic form. Study of the material allows us to focus on the action plan: Step 1—an idea of organization as a platform for the formation of a space of thinking and activity; the formation of platforms and specific phrases of the principles of organization of activity and ontological pictures and vision through them. There we use techniques presented in [34]. Step 2—the process of self-determination and schematization. The layer of thought activity, its formation and occupation, determination of the order of possible interactions, and communications, as well as reflective exit (mutation). The status of the scheme as the basis for determining the understanding of the texts of communication and capturing the meanings that the text carries on itself. Step 3—from positioning and sketching to objectification. Object as a result of the integration of self-determination, problematization, and schematization. An object as it is and a tool for the deployment of an organizational-activity plan. Step 4—from positioning and schematization through retrospection to problematization. Complex reflective transitions. Problematization is included in the text of thought-communication and serves as a basis for developing a picture of the world, the foundations of existence and individuation. Step 5—inverse processes. The impact of problematization on positioning and re-determination, the movement of a positional structure; the impact of problematization on the schematization and construction of tools that capture a thought from a communication text. Step 6—inverse processes. The impact of objectification on positioning and re-determination, the movement of a positional structure; the impact of objectification on the schematization and construction of tools that capture the idea from the text of communication. Step 7—integral view from the modus of the absolute. Relativity of the absolute, translation of culture and reproduction of activity. Norms and as the ultimate types of absolute and as moments that determine the principles of organization of activity and ontological design, respectively. Step 8—a bridge between self-determination in a positional structure (collective) and objectification and individuation.

If the substantive content is constantly kept in mind, and it is with it that we are obliged to constantly touch and shape it, then it is worth using different techniques.

If time is a decisive factor, then there is a middle ground between security and the speed of the cloud's response to a user's request. In other words, protection has its reasonable limit. And so that the attacker does not violate the integrity of the cloud, its normal functioning, a special kind of work is required with a potential client of this kind. Forcing, for example, to write complex programs for passing defense mechanisms, so that these codes can be used in crypto technologies.

4.2 Limitations

We limited ourselves to a fragment of an array of publications on the topic of cloud computing and ensuring their security. We did not conduct constructive criticism sufficient to decompose these texts into elements, units of a new assembly. The mention of a certain set of points that the authors of the articles draw attention to serves to approach the problem from different angles. This study is intended for practitioners who could better articulate their requirements for ordering a comprehensive methodological study.

We limited ourselves to take in account those works that will be made at Moscow Methodological Circle [34] concerning the systems and methodology [1].

5. Conclusions

An approach we have developed allows us to identify additional problems in this area and outline a program for their development. We try to build a system of

DOI: http://dx.doi.org/10.5772/intechopen.92143

methodological design and research over the many private methodologies that authors of articles usually use, relying on the experience of generalizing and concretizing system approaches, and, in particular, expanding geographical and historical boundaries, including system generalizations of intercultural studies and philosophical movements. An attempt is made to disassemble the security problem of cloud computing into a certain number of layers, processes, and technologies of thinking, and to reconnect them into a single whole with the character of thinking and activity.

The application of the methodological schemes of the general methodology allows us to transfer the body of texts of publications devoted to the security of cloud computing from the category of research and engineering to the category of practical, which would help to solve the problem of the relationship of openness of cloud environments and their protection from external and internal threats. We are strengthening the psychological thinking that underlies the agreement between the cloud computing provider and cloud users, design and research thinking based on substantive genetic logic. Its difference from formal logic is that its starting point is the situation that develops as a result of the functioning and development of a certain system of activity, in this case, the use of computing technologies in the cloud, the organization of this industry, and the provision of a normal functioning mode.

The expansion of the Internet of things with the inclusion of neuro prostheses [65] and nano mechanisms in this circle will give the methodological organization of security research a new meaning and additional significance. The transfer of the global economy (both at the planetary and local levels) to new platforms based on the inclusion of digital technologies in them will mean the isolation of the field of computing and the formation on the basis of cloudy and foggy computing of a sphere that needs proper immunity and its maintenance. The program idea of Society 5.0 will also require additional rethinking of the existing practice of protecting cloud computing from harmful influences

In our opinion, a systematic approach exists only as a unit and a particular organization of the approach "and the corresponding organization of thinking and activity" appear in the representatives of special sciences only because they borrow the means, methods and ontology of methodological methodology and methodological approach. The goal to combine several different objects could be achieved only by using the means and norms of methodology. The expression "system work," therefore, only describes the structure of methodological work and methodology; thus, we can approach the issue of the specifics of the system approach. If we choose a description in the theory of thinking, we will determine the specifics of systemic thinking. But a system approach can also be described in the means of the theory of activity, and then its specificity will be expressed and fixed differently. Thus, here too we must take into account the moment of multiplicity of possible representations. We have presented a figure in which we have reflected the principles of the methodological approach (**Figure 1(d)**), in which we tried to visualize methodological machine for creating the environment of successful decision of the Cloud Computing Security problems, listed in Section 2, "Literature review."

Conflict of interests

Authors have no conflict of their interests.

Author details

Svetlana Aristova[1], Yousef Ibrahim Daradkeh[2] and Petr Korolev[3]*

1 Lesinvest Co, Perm, Russian Federation

2 Department of Computer Engineering and Networks, Prince Sattam Bin Abdulaziz University, KSA

3 Studia Korolevae Int, Udmurt State University at Kudymkar, Perm, Russia

*Address all correspondence to: korolev@studiakorolevae.com

References

[1] Shchedrovitsky GP. Selected Works. Moscow: School of Cultural Politics; 1995. 800 p

[2] Trends in General Systems Theory. NY; 1972

[3] Dezani-Ciancaglini M, Montanari U. International symposium on programming. In: Proceedings of the 5th Colloquium (Turin, April 6–8, 1982). Berlin/Heidelberg: Springer; 1982

[4] Farnga M. Cloud Security Architecture and Implementation. A Practical Approach. Submitted to Prof Friedman. Towson University—Graduate School; 2018. p. 27

[5] Mohammad W, Das AK, Kumar N, Vasidakos AV. Design of secure key management and user identification scheme for fog computing services. Future Generation Computer Systems. 2019;**91**:475-492. DOI: 10.1016/j.future.2018.09.017x

[6] Guan Y, Shao J, Wei G, Xie M. Data security and privacy in fog computing. IEEE Network. 2018;**32**(5):106-111. DOI: 10.1109/MNET.2018.1700250

[7] Abdulrahman A, Deng Y, Wei G, Lin X. Collaborative security in vehicular cloud computing: A game theoretic view. IEEE Network. 2018; **32**(3):72-77. DOI: 1-.1109/MNET.2018.1700329

[8] Pinki S, Sengupta J, Suri PK. Survey of intrusion detection techniques and architectures in cloud computing. International Journal of High Performance Computing and Networking. 2019;**13**:2. Online publication date: 22 January 2019

[9] Ibrahim FAM, Hemayed EE. Trusted cloud computing architectures for infrastructure as a service: Survey and systematic literature review. Computers & Security. 2018;**82**:196-226

[10] Xu HJ, Zheng X. Security mechanism of dynamic and differentiated protection for telecommunications services based on cloud computing. International Journal of Security and Networks. 2018;**13**:4. Online publication date: 31 August 2018

[11] Maharajan K, Paramasivan B. Membrane computing inspired protocol to enhance security in cloud network. The Journal of Supercomputing. 2019; **75**(4):2181-2192. DOI: 10.1007/s11227-018-2629-6

[12] Meikang Q, Kung S-Y. Guest editor's introduction to the special issue on security and privacy on clouds. IEEE Transactions on Cloud Computing. 2018;**6**(2):301-302. DOI: 10.1109/TCC.2018.2790672

[13] Xu J, Liang C, Jain HK, Gu D. Openness and security in cloud computing services: assessment methods and investment strategies analysis. IEEE Access. 2019;7:29038-29050. DOI: 10.1109/ACCESS.2019.2900889

[14] Sajay KR, Babu SS, Vijayalakshmi Y. Enhancing the security of cloud data using hybrid encryption algorithm. Journal of Ambient Intelligence and Humanized Computing. 2019:1-10. DOI: 10.1007/s12652-019-01403-1

[15] Hang W, Hu G-Y, Han X, Qiao P, Zhou Z, Feng Z-C, et al. A new BRB Model for cloud security-state prediction based on the large-scale monitoring data. IEEE Access. 2017;**6**: 11907-11920. DOI: 10.1109/ACCEESS.2017.2779599

[16] Babin B, Zheng J. A Preliminary Study On Emerging Cloud Computing Security Challenges. ACM; 2018. DOI: 10.1145/1235

[17] De DM, Giaretta A, Dragoni N, Bucchiarone A. Cyber-Storms come

from Clouds: Security of Cloud Computing in the IoT Era. Future Internet. 2019;**11**:127. DOI: 10.3390/fi11060127

[18] Matheus T, Vieira M. Towards Models for Availability and Security Evaluation of Cloud computing with Moving Target Defence. Submitted on 3 September 2019 arXiv:1909.01392

[19] Pustovoi T. Recommendation systems of university 20.35 [Internet]. 2019. Available from: https://ntinews/blog/inside_outside/taras-pustovoy [Accessed: 27 November 2019]

[20] Gvishiani DM. Organization and Management. A Sociological Analysis of Western Theories. Moscow: Progress Publishers; 1972. 461 p

[21] Bogdanov A. General Organizational Science (Technology). Vols. 1 -3. 3rd ed. Moscow/Berlin; 1925/1929. [In Russian]

[22] Zadeh L, Desoer CA. Linear Systems Theory, the State Space Approach. New York: McGraw-Hill; 1963. 628 p

[23] Sadovsky VN. Foundations of General Systems Theory. Moscow: Nauka Publishers; 1974. 279 p. [In Russian]

[24] Mesarovic MD, Macko D, Takahara X. Theory of Hierarchical Multilevel Systems. New York/London: Academic Press; 1970. 294 p

[25] Uemov AI. Methods of construction and development of the general systems theory. In: Systems Research, Yearbook. Moscow: Nauka Publishers; 1971. pp. 146-178. [In Russian]

[26] Uemov AI. Systems Approach and General Systems Theory. Moscow: Mysl Publishers; 1978. 241 p. [In Russian]

[27] Views on General Systems Theory. New York: Wiley; 1964. 178 p

[28] Gnedenko BV et al. Large Systems: Theory, Methodology, Modelling. Moscow: Nauka Publishers; 1971. 289 p

[29] Engineering: Principles and Practice of Computer-based Systems Engineering. Chichester. John Wiley & Sons. ISBN: 0-471-93552-2

[30] von Neumann J. Probabilistic Logics and the Synthesis of Reliable Organisms from Unreliable Components. Automata Studies. 1956;**34**:43-98

[31] Herbert SA. The Sciences of the Artificial. Vol. 136. 3rd ed. The MIT Press; 1996

[32] Kallmann R, Falb PL, Arbib MA. Topics in Mathematical Systems Theory. New York: McGraw-Hill; 1969. p. 17

[33] Tabatchnikova S. Le Cercle de methodologique de Moscou (1954–1989): Unepensee, une pratique. Paris: Ecole des Hautes Etudes en Sciences Sociales; 2007. 332 p

[34] Klir GJ. Trends in General Systems Theory. New York: Wiley-Interscience; 1972. 462 p

[35] Warren W. Science and complexity. The American Scientist. 1948:536-544

[36] Steven W. A New Kind of Science. Wolfram Media; 2002

[37] Lofti Z. From circuit theory to system theory. Proceedings of the IRE. 1962;**50**(5):856-865

[38] Sadovsky VN. Some key problems in development of general systems theory. In: Systems Studies, Yearbook, 1971. Moscow: Nauka Publishers; 1972. pp. 35-54. [In Russian]

[39] Quade ES. Systems Analysis and Policy Planning. New York: Elsevier; 1968. 453 p

[40] Optner SL. Systems Analysis for Business and Problem Solving. New Jersey: Englewood Cliffs; 1965. 116 p

[41] Johnson F, Kast RF, Rosenzweig J. The Theory and Management of Systems. 2nd ed. New York/St. Louis/London/Sydney: McGraw-Hill Book Co; 1971

[42] Goode HH, Mackol RE. Systems Engineering. An introduction to the design of large-scale systems. In: Control Systems Engineering. New York: McGraw-Hill; 1962. 551 p

[43] Nikolayev VV. State-of-the-art and some problems of development of systems engineering. In: Methodological Problems of Systems Engineering. Leningrad: Sudostroyeniye Publishers; 1970. pp. 3-38. [In Russian]

[44] Simon H. The Sciences of the Artificial. Cambridge: MIT Press; 1969

[45] Laszlo E. Introduction to Systems Philosophy: Toward a New Paradigm of Contemporary Thought. New York: Gordon & Breach; 1972. 328 p

[46] Herbert SA. The architecture of complexity. Proceedings of the American Philosophical Society. 1962;**106**

[47] René T. Structural Stability and Morphogenesis: An Outline of a General Theory of Models. Massachusetts: Reading; 1972

[48] Schedrovitsky GP. Problems of Systems Methodology. Moscow: Znanije Publishers; 1964. 56 p. [In Russian]

[49] Schedrovitsky GP. On the characteristic of most abstract directions in methodology of systems–Structural studies. In: Problems of Research into Systems and Structures. Moscow: USSR Academy of Sciences; 1965. pp. 15-23. [In Russian]

[50] Schedrovitsky GP, Yudin EG, Lefebvre VA. The 'natural' and the 'artificial' in semiotic systems. In: Semiotics and Oriental Languages. Moscow: Nauka Publishers; 1967. pp. 48-56. [In Russian]

[51] Schedrovitsky GP. Methodological meaning of linguistic universals. In: Linguistic Universals and Linguistic Typology. Moscow: Nauka Publishers; 1969. pp. 46-98. [In Russian]

[52] Spirkin AG, Sazonov BV. Reflection on methodological problems of research into structures and systems. Voproey Filosofiji. 1964;**1**:15-47. [In Russian]

[53] Dubrovsky VY, Shchedrovitsky LP. System Approach to Human Factors Engineering. Moscow: University Press; 1971

[54] Guschin YF, Dubrovsky VY, Schedrovitsky LP. On the concept of systems design. In: Large Information Control Systems. Moscow: Moscow House of Scientific and Technological Education; 1969. 82 p. [In Russian]

[55] Kuzmin VP. Systems Principle in Theory and Methodology of Karl Marx. Moscow: Politizdat Publishers; 1976. 261 p. [In Russian]

[56] Development and Implementation of CAD (Theory and Methodology). Moscow: Stroyizdat Publishers; 1975. 527 p. [In Russian]

[57] von Bertalanffy L. Modern Theories of Development: An Introduction to Theoretical Biology. New York: Oxford University Press; 1933

[58] von Bertalanffy L. General System Theory: Foundations, Development, Applications. New York: George Braziller Inc; 1968

[59] System Theory [Internet]. Available from: http:/en.wikipedia.org/windex.

php?title=Systems_theory&oldid=
935325354 [Accessed: 06 March 2020]

[60] Dubrovsky VY. Three System Paradigms (Plato, Aristotle, Schedrovitsky). Kudymkar: Studia Korolevae Int; 2020. [In Russian]

[61] Zilberman DB. Genesis of Meaning in Hindu Philosophy. Moscow: Editorial URSS; 1998. p. 448. [In Russian]

[62] Ackoff RL. A Concept of Corporate Planning. New York: Wiley-Interscience; 1970. 158 p

[63] Daradkeh YI, Aristova SM, Korolev PM. Observation and Audit of the Processes in Experiences with Uncertainty. Journal of Computer Engineering & Information Technology. 2016;5:4. DOI: 10.4172/2324-9307.1000163

[64] Aristova S, Korolev P. Knowledge-information transformation: Reflexive games on human language. In: Proceedings of International Symposium on Reflexive Control; 17-19 October 2000; Moscow. Moscow: RAS Institute on Psychology; 2000. pp. 61-62. [In Russian]

[65] George Braziller Inc. General System theory: Foundations, Development, Applications. Revised edition 1976. New York: George Braziller Inc. 1968. ISBN: 0-8076-0453-4

Chapter 4

Security at the Edge

Charles J. Gillan and George Karakonstantis

Abstract

The Internet has become an essential part of daily life for almost everyone in society having grown far beyond its roots in the 1970s as the ARPANET, a network that was principally the domain of scientists and engineers. The popularity of the HTTP, developed at CERN in the late 1980s led to the widespread use of the term 'the web' as a generic name for the Internet for many years, at least in the public domain. Of course, the Internet is much more than just web browsing and, in recent years, the term cyberspace has become the most popular term to describe interactions over the Internet. Yet, an unambiguous definition of the term is difficult to formulate . Financial institutions underpinning the economy and the operation of national critical infrastructures, such as monitoring and control of the electricity supply, are now dependent on the Internet. A consequence of this is that cyberattacks become more costly for the victims and perversely more attractive to the criminals who carry them out . The advent of the Internet of Things (IoT) and edge computing as a new paradigm creates the potential for enhanced productivity but at the same time opens up new opportunities for cyberattacks while still being exposed to existing attach vectors such as the well-known denial of service attack (DDoS), which can take place in many forms . In this chapter, we described the challenges in building an edge system that is secure against cyberattack. We begin by briefly reviewing the architecture of communications over the Internet and later consider the new challenges that follow from operating the hardware with values of voltage, frequency and current that enable more energy efficiency.

Keywords: security, energy efficiency, performance, cloud, edge computing

1. Introduction

The Internet has become an essential part of daily life for almost everyone in society having grown far beyond its roots in the 1970s as the ARPANET, a network that was principally the domain of scientists and engineers. The popularity of the HTTP, developed at CERN in the late 1980s, led to the widespread use of the term 'the web' as a generic name for the Internet for many years, at least in the public domain. Of course, the Internet is much more than just web browsing and, in recent years, the term cyberspace has become the most popular term to describe interactions over the Internet. Yet, an unambiguous definition of the term is difficult to formulate [1].

Financial institutions underpinning the economy and the operation of national critical infrastructures, such as monitoring and control of the electricity supply, are now dependent on the Internet. A consequence of this is that cyberattacks become more costly for the victims and perversely more attractive to the criminals who carry them out [2]. The advent of the Internet of Things (IoT) and edge computing as a new paradigm creates the potential for enhanced productivity but at the same time opens up new opportunities for cyberattacks while still being exposed to existing attack vectors such as the well-known denial of service attack (DDoS), which can take place in many forms [3].

In this chapter, we described the challenges in building an edge system that is secure against cyberattack. We begin by briefly reviewing the architecture of communications over the Internet and later consider the new challenges that follow from operating the hardware with values of voltage, frequency and current that enable more energy efficiency.

2. The structure of the internet: security from data Centre to the edge

There is a proverb in the English language that says that a chain is only as strong as its weakest link. This applies directly as a basic principle of cybersecurity. Edge computing still requires communications to a central data centre, at least some of the time. It follows that it is necessary to consider carefully the WAN and LAN technologies used. **Figure 1** illustrates the networking technologies used and shows the position of edge computing within the wider fog computing environment, which we describe in a later part of this section. The section begins by discussing each networking technology separately and in doing so refers briefly to the history of the development of data networking technologies in general and to the development of the Internet in particular.

2.1 WAN technologies—circuit-based communications

The core transmission technology of the global telephone system developed over several decades from using electromechanical switches and frequency division multiplexing to use digital signals and time division multiplexing by the 1980s. Signals from different sources were multiplexed together in a hierarchy of data rates (2.048 Mbps, 8.448 Mbs, 34.368 Mbps, etc.) for transport across the core network

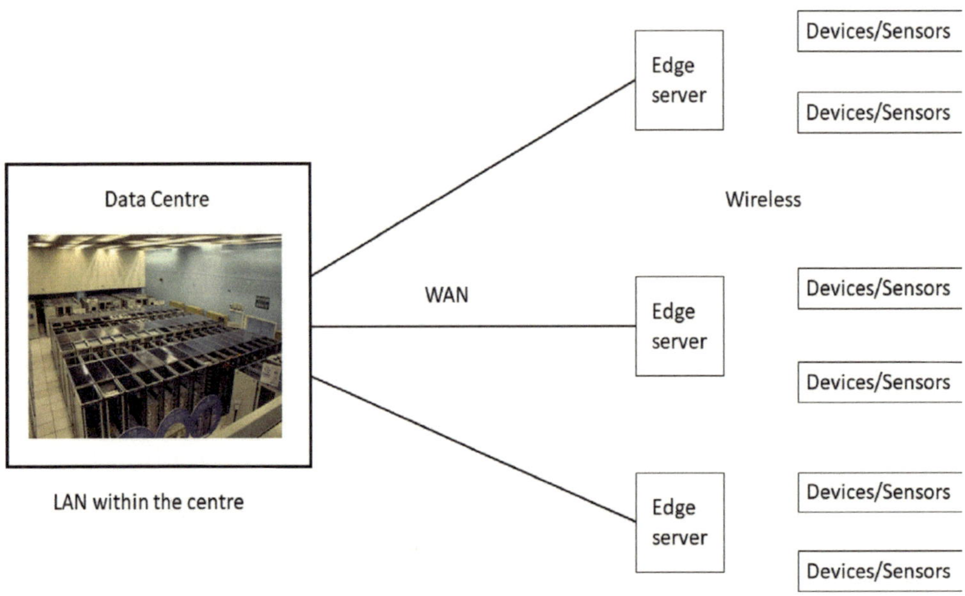

Figure 1.
Illustration of the hierarchy of devices creating the fog computing environment.

before being demultiplexed for transmission to individual receivers. The concept of a unique end-to-end circuit from sender remained clearly identifiable.

The initial plesiochronous digital hierarchy (PDH) handled the fact that lower bit rate sources were not time synchronised (each source had its own clock) by adopting the technique of bit stuffing in order to ensure that the higher rate channels were time synchronised. Thus, equipment inserted extra bits, as needed, at the transmitter and then the receiver removed these bits.

As fibre optic became widely used in the telecommunications industry, PDH was replaced by a different, more scalable, multiplexing technology known as the synchronous digital hierarchy (SDH) in which the equipment across the network is synchronised. SDH works on copper lines and on radio signals as well as fibre optic cables. The ITU-T [4] develops standards for SDH globally. The United States developed the technology under the name Synchronous Optical Network (SONET) around the same time as the ITU-T. In SDH, an aggregate signal composed of virtual containers (VCs) of fixed size is transmitted at a fixed frequency between two pieces of SDH equipment. Each tributary signal arriving at the sender from a source is assigned to one of the VCs with a pointer indicating where the signal is located within the container. Thus by allowing the pointer to vary, the tributary signals are adapted to the synchronised clock of the transmitter and receiver.

While a transmission from source to receiver will pass though many different VCs as it transits the SDH network, essentially using a different VC on each point to point link, the concept of an identifiable circuit remains intact in SDH. This means that distinct users and applications are clearly separated despite the fact that they are carried over the same fibre, wire or radio link. Even if one captures the complete SDH aggregate signal, without knowledge of the mapping of users and applications to the VCs in the signal, it is essentially impossible to extract the targeted data stream.

2.2 Packet communications

The circuit concept in the telephone system described in the above section builds on the idea of reserving bandwidth between the transmitter and receiver although as we have mentioned this confers a certain level of security by separating the signal from others on the same physical medium.

An alternative approach that is available when the transmission is in digital form is to break it into parts and then to transmit these parts in sequence across the digital network. We can define a packet to have three parts: a header, a payload and optionally a trailer. Each part of the digital data is placed uniquely into one packet and the header defines the information that allows the packet to be transmitted across the digital network. This type of transmission, known as packet switching, is the primary basis for data communications in computer networks, whether local or wide area. The definition of the fields in the header (and trailer, if present) plus the functionality associated with each field defines a protocol. The development of early networks, such as the ARPANET discovered that it was useful to encapsulate protocols within other protocols leading to the concept of a layered stack. This was eventually formalised in the definition of a seven-layer abstract model known as the Open Systems Interconnection (OSI) model [5].

As the Internet was adopted globally in the 1990s, intense efforts were applied to use the existing global SDH network, as the wide area networking technology (WAN), to carry the packet protocols that underpin the physical layers of the Internet. Packet over SONET (POS) was developed, defined in RFC 2615 [6] initially, as a way of transmitting packet-based data protocols using point to point protocol (PPP) on each point to point link in an SDH/SONET network.

POS includes the option to apply scrambling to the transmission thereby adding an extra layer of security.

2.3 From cloud to edge to fog computing

The global adoption of the Internet enabled cloud computing paradigm. Large data centres, using virtualisation technology, can offer end users scalable compute resource on a pay per use basis. This approach is well suited to traditional enterprise computing freeing businesses from capital expenditure on computing systems transferring the cost to operational expenditure and off-loading risk to cloud service providers.

Newer applications, such as the Internet of Things (IoT), involve data collection at end user devices, equipment that is often mobile and therefore linked by wireless to edge nodes. The complete system is geographically diverse, with Smart Cities being one of the best illustrations of this being. The opportunity to redistribute computation across the hierarchy from user device, through edge, and back to the data centre when needed is now called fog computing. A hypothetical IoT service with a target end-to-end latency of 200 ms can easily expect, for a roundtrip to the cloud, to spend half of its budget in the network. This leaves a very tight time budget for execution of the actual processing to at the data centre. Fog has the potential to eliminate most, if not all, of the communication latency and, therefore, can permit the option of running the edge systems at lower frequency and voltage; for example, operating at 50% of the peak frequency with 30% less voltage translates to running with 50% less energy and 75% less power. Edge servers can also benefit from virtualisation, running multiple virtual machines to separate functionality. Furthermore research suggests that compute accelerators, in particular GPUs, may be enabled at the edge though virtualisation [7].

Figure 2 shows an analysis of the operation of an edge server, operating in extended margins, presented by the Horizion 2020 project Uniserver (http://www.uniserver2020.eu). UniServer created a cross layer approach from the hardware levels up to the system software layers. The following system enhancements were identified:

i. **at the circuit, micro-architecture and architecture layer** by automatically revealing the possible operating points (i.e. voltage, frequency) of each hardware component no worse than the worst-case operating points used, thus helping to boost performance or energy efficiency at levels closer to the Pareto front maximising the returns from technology scaling;

ii. **at the firmware layer** with low-level handlers by monitoring and controlling the operating status of the underlying hardware components and updating a 'HealthLog', as well as performing periodical benchmarking of the hardware and reporting the findings in a 'StressLog'. The logs with the collected information are communicated to the software stack (hypervisor) in a generic way, allowing easy adoption and exploitation of the observed margins;

iii. **at the software layer** by enabling an easy programmability, ensuring high dependability and full utilisation of the margins observed in the underlying hardware. State-of-the-art software packages for virtualisation (i.e. KVM) and resource management (i.e. OpenStack) will be ported on the micro-server further strengthening its advantages with minimum intrusion and easy adoption.

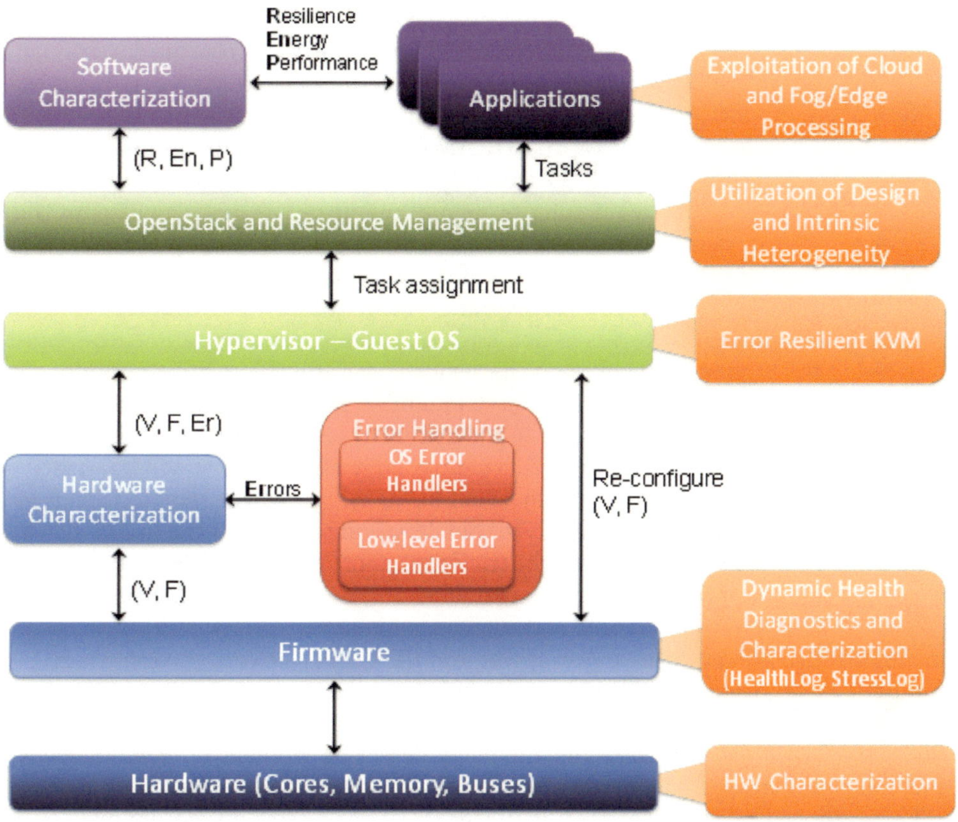

Figure 2.
Perspective from the UniServer project on the enhancement of the edge server.

In this chapter, we focus on the security challenges at the edge components and as we have outlined the WANs that link these edge nodes back to the data centres.

3. Cyber security at the edge

The edge computing paradigm moves significant amounts of computation from the data centre closer to the source of the data, reducing but not eliminating the need for packet communications. It follows that there are larger number of smaller servers deployed at the edge and therefore energy efficiency of the server operation becomes a significant factor. Edge servers have fewer CPUs and less DRAM and limited power budgets when compared to rack mounted servers in data centres. One driver for this is often the fact that physical form factor of the edge server is significantly limited compared to rack space in the data centre.

Manufacturers of server components define operational limits for parameters such as voltage, frequency and current. Routine adherence to these limits in the production of commercial servers reflects, in part, the need to account for the expected performance degradation of transistors and potential functionality failures due to the increased transistor variability in nanometre technologies. In general, the values adopted are quite pessimistic. DRAM manufacturers, in an effort to limit the potential faults, adopt a high operating supply voltage and refresh rate according to the assumed rare worst-case conditions [7]. This leads to the observation that DRAM alone can account for up to 40% power usage. Researchers have however investigated the operation of these electrical components in regions of voltage and current beyond the conservative limits [8] and report 8.6% system energy savings on average for non-virtualised and 8.4% for virtualised workloads while ensuring the seamless server operation even under extreme temperatures.

Relaxation of voltage, timing and refresh-rate limitations may put at risk the correct functionality of the CPUs and DRAMs due to the potential failures that may occur at lower voltages and dynamically changing operating/environmental conditions (e.g. temperature). Such timing and memory failures may disrupt the operation of the server and/or directly impact the expected Quality of Service (QoS), which can be quantified in terms of throughput and quality-of-results (e.g. in terms of Bit-Error-Rate). As a consequence, such failures will affect service level agreements (SLA) in terms of availability, latency, accuracy and throughput as agreed at the higher level between the service user and the service provider. A further consequence of operating in these extended margins is that new security vulnerabilities may arise in addition to the cyber threats that already exist.

In contrast to a centralised cloud data centre, edge deployments will be constituted from many small clusters or individual installations, where elevated levels of physical security are not economically viable. Physical security of the micro-server may consist primarily of a light-weight enclosure and, from a security perspective, it should be assumed that a determined attacker will be able to gain full access to the system. This creates a larger threat surface, which now incorporates physical attacks, posing threats to the micro-server and the wider network it connects to. Deployments at the edge should be made under the assumption that networks are operating over untrustworthy links, with the use of encrypted tunnelling through VPNs, malware detection, firewalls, intrusion detection/prevention systems and DNSSEC all considerations for an endpoint security policy.

Threats posed by attackers gaining physical access to a system require consideration from both hardware and software security disciplines. Applications developers should employ secure coding practises, particularly when operating on any sensitive information. Care should also be taken to minimise, or, if possible, to avoid the storage of secret information in physical memory. The use of software, or ideally hardware-based, hard disk encryption technologies can offer protections, even when the disk is removed from a system.

Side-channel attacks can potentially be used to reveal sensitive information. In the UniServer system, sensitive extended margin information could be targeted to create denial of service attacks or cause system instability. The variation of voltage and frequency margins, core features of the UniServer solution, may also influence the relative amount of side-channel leakages. Side-channel resilient countermeasures, employing masking and hiding strategies, should be employed to help counteract such threats.

The differing deployment architectures of full stack and bare metal are considered. In the full stack deployment, representing a micro-server data centre, the UniServer software is running under the host OS, abstracted from other guest applications under separate virtual machines. However, in the bare metal deployment, the UniServer software runs along-side other system applications. It is in this deployment architecture where the UniServer system is most exposed to interference by other applications. The UniServer log files are identified as high value assets that need to be protected from tampering, since it could potentially lead to system instability or denial of service attacks. It is therefore a recommendation that the log and policy files are stored in an encrypted format, to avoid reading and manipulation by others. Additionally, consideration should be given as to whether the files should be digitally signed, to provide assurance that they come from a trusted source. These recommendations would naturally have overheads in terms of real-time operation, so their implementation would need to be considered carefully in terms of system performance. The use of encryption, and possibly digital signing, will likely be candidate to form a security solution.

3.1 General attack vectors

In this section we consider the threats posed to both traditional networked server infrastructure and to the class of physical attacks, discussing the threats and countermeasures used to mitigate against them.

The primary aims of information security are to ensure the confidentiality, integrity and availability of a system [9]. There is generally no single solution to a security problem, since threats and vulnerabilities originate from many sources; rather the aim is to provide a series-layered security response, delivering defence in depth. An overall security response should be considered in the wider sense, consisting of measures that span the range of administrative, logical/technical and physical solutions.

3.1.1 Security of the operating system

The operating system (OS) is the fundamental software layer upon which the rest of the system software is built. In the common four-ring model, shown in **Figure 3**, the operating system is separated into two distinct regions of Kernel space, incorporating kernel memory, components and drivers from rings 0 to 2, and user space in ring 3, where end user applications may be run.

For most commercial operating systems, control of user access is organised under discretionary access control (DAC), providing privileges at the individual user account level. However, unlike a system under mandatory access control (MAC), where applications run in isolated memory with strong separation, typical OS's are running in a multi-tasking environment where resources are shared and are potentially accessible between applications [10]. Security is, therefore, ultimately left up to the system administrator to ensure that appropriate measures are in place and that the system is configured appropriately. Some general recommendations for operating system security, which apply to both cloud and edge deployments, are summarised below [11].

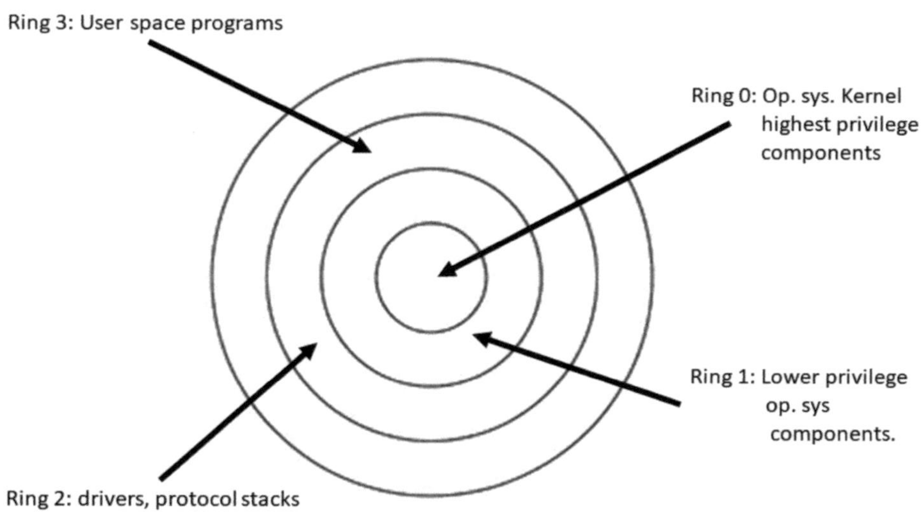

Figure 3.
Layers of protection in the operating system.

3.1.2 System integrity

- Build production systems from a known and repeatable process to ensure system integrity.
- Check systems periodically against snapshots of the original system.
- Use available third-party auditing software to check system integrity.
- Backup system resources on a regular basis.

3.1.3 User accounts

- Limit the number of user accounts.
- Ensure that only a few trusted users have administrative access.
- Assign the minimum required access permissions for the account that runs an application.

3.1.4 Password policies

- Require the use of secure passwords, that is, passwords of sufficient length, using a mix of letters, numbers and symbols. Do not re-use passwords and avoid the use of any personal information or dictionary words.
- Use automated tools to try and crack any weak passwords and require their update by users.
- On a UNIX operating system, activate the shadow password file.
- Use two-factor authentication.

3.1.5 File system

- Deny access by default.
- Provide minimal access rights where necessary, for example, read only.

3.1.6 Network services

- Provide the minimum number of required services.
- Reduce the level of access permissions for network services users.
- Ensure that user accounts that have access to the Web server do not have access to shell functions.
- For UNIX/Linux, ensure that unused services do not exist in the rc files, rc0-rc6, in the /etc. directory.
- Ensure that unused services are not running, and that they do not start automatically on MS Windows.

- Reduce the number of trusted ports specified in the /etc./services file.
- Protect your system against NetBIOS threats associated with ports 137, 138 and 139.
- Use wrapper services, such as iptables.
- Avoid using services that have a GUI, since such services introduce many known vulnerabilities.

3.1.7 System patches

- Run the latest, vendor-recommended patches for the operating system.
- Schedule regular maintenance of security patches.

3.1.8 Operating system minimisation

- Remove non-essential applications to reduce possible system vulnerabilities.
- Restrict local services to those required for operation.
- Implement protection for buffer overflow.

3.1.9 Logging and monitoring

- Log security-related events, including successful and failed logons, logoffs and changes to user permissions.
- Monitor system log files.
- Use a time server to correlate time for forensics.
- Secure the system log files by restricting access permissions to them.
- Secure the logging configuration file.
- Consider the use of a remote server for storage of logging information.
- Enable logging of access requests on web servers.

3.1.10 Hyperjacking

Hypervisor technology enables the deployment of numerous virtual machines (VMs) on the one system, indeed it is a key concept in shared cloud infrastructure. However, the deployment of multiple systems adds complexity and consequently the possibility for new exploits. The term virtualisation escape, or VMEscape, refers to the process by which an attacker can escape the confines of the virtual environment and is then able to exploit the host OS. Virtualised systems should therefore still be deployed under the supervision of firewalls, while guests with differing security levels, such as DMZ and internal, should not be combined on the same host.

It has been reported that malware rootkits have also been developed that act as hypervisors, installing themselves below operating systems, in a process referred to as hyperjacking. Since this software operates ostensibly outside the scope of the operating system, it can evade malware scans and also spy on the system, gathering information such as logging of passwords. In 2009, researchers from Microsoft and North Carolina State University revealed Hooksafe [12], a hypervisor class anti-rootkit, aiming to demonstrate the provision of generic protection against kernel-mode rootkits.

3.1.11 Network attacks

Access via network ports forms the basis of most remote attacks on cloud-based infrastructure. The ports of machines around the world are continually being probed to see if any ports have been left open or unsecured. It is therefore a basic preventative measure to close any unused ports and restrict access and secure those essential ports that are required to remain open. Improperly implemented TCP/IP stacks are vulnerable to various attacks such as buffer overflows, SYN flood attacks, denial of service attacks such as Smurf, ping and Fraggle and fragment attacks such as Teardrop to name but a few. These attacks can be largely mitigated by applying the appropriate configuration to disable services and apply the relevant patches.

Under the assumption that edge-deployed servers are more exposed, there are numerous means by which the traditional networking security elements of fire-walls, proxies, virus scanners can be circumvented, creating a means by which other nodes of the network may be exposed. In 2014, the Gameover Zeus (GOZ) botnet was responsible for the global distribution of the CryptoLocker ransomware, which encrypted the victim's hard drive and required payment to receive the decryption key.

Since network connections could be exposed, the communications channel of an edge device should be considered untrustworthy, since attacks such as eavesdropping on network traffic, man-in-the-middle, modification or replay attacks are all possible. It is recommended that an encrypted VPN tunnel should be used between the edge server and other elements of the network to mitigate against such attacks.

DNS hijacking exploits the vulnerability in the way local or caching DNS servers obtain information from root servers regarding the identity of the authoritative servers for a domain. It is possible for an attacker to send falsified replies, and thus control the domain resolution, forwarding the user to the attacker's server [13]. The most effective countermeasure against DNS hijacking is to upgrade DNS to Domain Name System Security Extensions (DNSSEC).

When considering the above attacks, it is evident that edge deployments should incorporate their own endpoint security, consisting of elements such as inbound/outbound firewalls, malware scanning and intrusion detection/prevention systems as necessary security countermeasures.

3.2 Physical attacks and countermeasures for edge deployments

We now turn our attention to the situation in which a determined attacker has been able to bypass the limited protections of an enclosure and has gained direct physical access to the system, providing an enhanced ability to tamper with the system. There are many such physical attacks referenced in the literature; here we aim to give an overview of attacks, providing examples for the most relevant and practical attacks, along with examples of suggested countermeasures to those attacks.

3.2.1 Memory attacks

High-performance, processor-based, systems will generally include the following types of memory: L1/L2/L3 cache, DRAM, Flash Firmware and Hard-Disk Drives. Each of these is a potential threat vector for an attacker.

3.2.2 Timing attacks

Timing attacks exploit the differences in time required to perform specific operations. For example, the time required to calculate division and multiplication instructions, or the time necessary to fetch data when a cache hit, or cache miss, is experienced. Similarly, the difference in timings when conditional branching is used, or when optimisations are used by a programmer to skip unnecessary operations, may improve application performance but at the same time can reveal sensitive information about underlying code and values being processed. A classic example was shown by Kocher in [14] where the timings for modular multiply operations in exponentiation operations, and modulo reductions of the Chinese Remainder Theorem (CRT) optimisation in RSA, could lead to the discovery of the entire encryption key on a PC.

An example of a remote network-based attack is that of Bernstein in [15], demonstrating a timing attack on OpenSSL AES, on a UNIX x86 server. The server was profiled using a known key to determine the timing characteristics for the input plaintext values. During the attack, plaintexts were sent to the server, with their timing profiles compared to the profiled reference. The information leakage was reported to be due to the non-constant timing of table lookups.

Cache-timing attacks were first proposed by Page in [16] and demonstrated by Tsunoo et al. in [17], where DES was broken with a > 90% success rate. In [18], Tromer et al. showed that the full AES key could be extracted using DM-CRYPT disk encryption on Linux with only 800 accesses to an encrypted file. The attack took 65 ms of measurement time and 3 seconds to analyse. The OpenSSL library was also attacked in as little as 13 ms, with 300 encryptions.

Countermeasures to timing attacks generally aim to perform operations in constant time. However, this is not a straight-forward task since compilers can often provide optimisations that affect timing behaviour. In addition, cache hits and variances in instruction timings are generally outside the control of the software designer. A clock-skipping countermeasure was initially proposed by Kocher in [19], which inserted random delays to try and break up characteristic timing patterns, but this was later shown to be equivalent to adding noise to the power waveforms and could be overcome by analysis with a larger number of traces.

In [18], Tromer et al. considered various countermeasures against cache attacks. They suggested:

1. Avoid the use of memory accesses by replacing lookups with equivalent logical operations. This is a possibility for algorithms such as AES. However, there will be a performance trade-off.

2. Use of a bit-slicing approach.

3. Use of a cache no-fill mode, where memory is accessed from the cache during a hit and serviced from memory when there is a cache miss.

4. Dynamic table storage, where the contents of the table lookup are cycled around in memory during encryption operations to de-correlate it.

Guidance for coding standards for cryptographic implementations in software can be found in [20]. For example, in the context of timing attacks, it is recommended:

1. Do not compare secret values on a byte-by-byte basis.

2. Avoid branching predicated on secret data.

3. Avoid the use of lookup tables indexed by secret data.

4. Avoid loops that are bounded by a secret value.

The software developer can also make use of libraries, written with security in mind, such as NaCl [21] and some processors also include custom instruction sets dedicated to cryptography, such as the Intel AES-NI instructions referenced in [22] and the ARM cryptography extensions discussed in ARMv8 [23].

3.2.3 DRAM attacks

Buffer Overflow is a well-known attack that can enable execution of malicious code. Strategies to counteract this attack include the use of improved input validation and bounds checking at the programmer level, or at the system level through approaches such as the randomisation of memory layout or the structuring of buffer memory to incorporate memory spaces, sometimes termed 'canaries', that actively monitor to detect when unauthorised overflows occur.

The purposeful use of errors, exceptions and crashes can also be used to initiate memory dumping, where the entire contents of system memory are exported to enable readout of sensitive values stored in memory. It is recommended that sensitive values should not be stored in memory in the clear, rather they should be stored in encrypted form, or represented as hashed values and compared against re-computed hashes when required.

With direct physical access to a system, such as with an exposed and isolated edge server, an attacker can potentially remove DIMM memory modules from the system board. As described in [24], the use of cooling sprays can enable a DIMM memory module to retain memory, without error, for several minutes. The memory can then be plugged into another system and sensitive information read out. This attack has been shown to make on-the-fly software-based disk encryption systems such as BitLocker, FileVault and TrueCrypt vulnerable. One countermeasure approach would be to avoid the use of pre-computed tables of information for encryption routines, which would typically be stored in DRAM, although this will have performance penalties associated with it since the values will need to be computed on-demand each time.

RowHammer is a more recent memory attack that exploits a weakness identified in commodity DRAMs, where repeated row activations can cause bits to flip in adjacent rows. A recent attack [25] used generic memory functions such as libc, memset and memcpy for attack primitives, making the attack more accessible.

3.2.4 Re-flashing attacks

Re-flash attacks target the replacement of existing system firmware with that of compromised firmware images. This can enable attackers to circumvent protections that would otherwise be in place. Due to the low-level nature of firmware access and control, such attacks can have a powerful effect on a system. Countermeasures may include incorporating password access for flashing operations.

3.2.5 Hard disk drive attacks

Hard drives will generally host the main operating system and the application software that loads on the system, but also potentially swap page information, which may hold sensitive information temporarily stored from primary DRAM memory. Hard disks, and particularly hot-swappable server-class drives, can be removed from a system at ease, and then connected to another system by plugging in a power and data cable. The disks can then be mounted as secondary drives to be copied, interrogated, or have additional malware or software installed. All of this is outside the scope of any protection from intrusion prevention systems of the original host. It is therefore advisable to consider the deployment of disk encryption technologies, such as software-based encryption, or preferably, hardware-based total disk encryption.

3.2.6 Side-channel attacks

We now consider a class of physical attacks termed as side-channel attacks. These attacks target the leakage of information from a system and are primarily concerned with the discovery of the secret information such as encryption keys that underpins modern cryptographic processing. The same approach can be targeted at modelled leakages of any other high-value information that is processed in a system.

3.2.7 Power analysis attacks

Power analysis is a powerful technique used to obtain side-channel information from a system. The power analysis attack can be categorised into two types: simple power analysis and differential power analysis.

In simple power analysis, the individual power waveform acquisitions are observed to see if information can be gleaned from them. In the attack of [14], it was observed that a single power consumption trace could reveal the entire encryption key by simply interpreting the pattern of the power trace, since modular multiply operations in exponentiation operations took varying times depending on whether the portion of the encryption key was a '1' or a '0'.

In differential power analysis (DPA), a series of power consumption measurements are recorded while the device is processing the target information, typically a secret encryption key, and is then compared against a set of hypothesised power models to determine a portion of the key. The analysis is repeated for the remainder of the key portions until the complete encryption key is recovered, enabling the attacker to decrypt any data, previously encrypted with the same key. Power consumption is typically modelled by estimating the number of '1's in a register via a Hamming weight or Hamming distance power model. Several differing methods of statistically comparing the modelled versus measured power consumptions are commonly used, such as difference of means, distance of means and Pearson's correlation coefficient [26].

Power analysis attacks are device specific and it can take from several hundred, to several million, traces to break an implementation with a DPA attack; this is dependent on the signal/noise (S/N) ratio and whether any countermeasures are present. Research has been carried out on a multitude of low-frequency embedded systems, where the approach has proved very successful. The attack works best when a clean voltage signal is available, preferably from the processor core of the device, where S/N is typically optimal; however, attacks can also be mounted by measuring the global power supply of a device through the voltage drop across a small resistor placed between supply and ground. There are fewer published works that address attacks on full-scale server boards, due to the additional complexities introduced by higher frequencies of operation, lack of access to processor core voltage and the additional noise generated by numerous system hardware elements.

Countermeasures against power analysis attacks aim to break the statistical link between the power consumption and the sensitive intermediate data values. For defence against simple power analysis, countermeasures primarily focus on disturbing the power waveform to disrupt the observable pattern, and so remove the discernible information. This can be accomplished by increasing background noise signals, introducing random insertions or delays, or by removing conditional branching and employing constant time algorithms.

Protecting a device from DPA is a much more challenging task, since this attack uses advanced statistical techniques to extract information from many traces. Countermeasures can be classed into two broad categories, namely whether they aim to hide or mask the data [27]. Hiding approaches do not attempt to change the intermediate values that are processed, rather they try to change the power waveform by applying some randomisation or by making it constant. Randomising approaches were mentioned above for simple power analysis measures and could also include approaches such as shuffling or skipping of instruction clocks. To make the power consumption constant, approaches have been proposed such as the use of dual-rail pre-charge (DRP) logic styles, which uses two wires that are complementary for each signal. Other logic styles, such as Sense Amplifier Balanced Logic (SABL), were proposed by Tiri et al. in [28] to provide resistance against DPA. However, these approaches require custom ASIC design with careful layout considerations and have still been shown to be vulnerable to DPA attacks.

The masking countermeasure aims to change the sensitive intermediate values by applying and then removing a temporary mask operation, a simple example being an XOR with a random value. This then breaks the link between what the power model expects and what is processed inside the device. The disadvantage of masking is that it can require the application and removal of multiple masks, for example switching between Boolean and multiplicative masks. This has a processing overhead and can be complicated to design and implement.

3.2.7.1 Electro-magnetic attacks

Electro-magnetic (EM) attacks [29] are a variation of power analysis attacks. They differ in the method of acquisition, which uses an electric or magnetic field probe to convert EM radiation into voltage signals that are proportional to the power consumption. The probing is generally classed as being either near-field or far-field. Near-field probing is considered to be the short-range distance that is typically less than one-wavelength from the source. At this distance, the field strength is proportional to $1/r^3$ in strength, therefore placing the probe as close as possible to the source will maximise signal strength. A more invasive attack can be to remove the chip package surface and enable a fine point-tip probe to be placed very close to the exposed integrated circuit (IC); however, this requires more time and generally a laboratory environment. A less invasive approach is to rest a simple loop antenna or EM probe tip against the surface of the IC, and to use active amplification to improve signal strength for appropriate quantisation scaling during acquisition.

Far-field EM attacks work at multiple wavelength distances and typically use a high-frequency directional antenna to receive signals. The waveforms being captured here have escaped the confines of the near field and are propagating over free space [30]. This form of attack would likely only be possible for exposed, non-shielded enclosures.

An EM acquisition can have advantages over that of traditional power analysis attacks. Firstly, it can have a lower invasiveness. In comparison to a power analysis attack, where a resistor may need to be soldered into place, the EM probe can often be placed in close proximity, without any evidence of tampering. Secondly, there is the possibility to improve the localisation of the probe, that is, to position it directly around the circuitry processing the sensitive information. This can help reduce the contributions of the EM fields generated from other elements of the overall power consumption. This can improve the S/N ratio, making it easier to visually identify leakages on an oscilloscope and improves the statistical analysis.

The countermeasures of hiding and masking, discussed above, also provide general protection against both EM analysis. However, for non-invasive attacks with an EM probe, physical shielding countermeasures can offer some further resistance. In [31], Yamaguchi *et al.* applied thin magnetic film to shield an integrated circuit device and reported a 6 dB reduction in magnetic field signal strength.

3.2.7.2 Profiling attacks

Profiling, or template, attacks [32, 33] use a reference device to build a characteristic power model of a device for various test inputs. The power model can then be compared against the power consumption measurements of an identical device to reveal what data have been processed internally. The template attack can potentially reveal the secret key with as little as one power trace; however, to obtain a power model with high fidelity may require the acquisition and pre-processing of many power traces, which may be a time-consuming exercise. Masking or the randomisation of execution order could be used as potential countermeasures.

3.2.7.3 Machine learning attacks

Machine learning is an emerging approach to side-channel attacks. Although numerous algorithms can potentially be used, the specific feature selection and data set size have the major influence on the success of the attack. Examples of approaches are supervised learning, support vector machines, random forest, neural networks and unsupervised learning. To date, most research has focussed on support vector machines [34–36], random forest [37] and neural networks [38]. Countermeasures to machine learning may include higher order masking approaches and the use of poisoned data.

3.2.8 Fault attacks

Fault attacks aim to induce erroneous behaviour in devices by inserting transient faults that propagate through the system and reveal secret information as a consequence. The transient nature of the targeted faults means that an attack can be attempted repeatedly, and the attack developed. This approach means that no permanent damage is caused to the device and therefore it is less likely that any evidence remains that an attack has taken place. In [39, 40] it was shown that faults could be induced in smart card devices by varying the system supply voltage, clock speed and ambient temperatures. Since these same characteristics are altered in

UniServer, it is an area of active investigation in the project, for example in terms of generation of memory and system errors.

Fault attacks in the literature have targeted both public and private key algorithms. Consider, for example, the attack on the Chinese reminder theorem (CRT) computation in RSA of [41] and the targeting of AES in [42, 43]. The attack of [43] demonstrating that inducing two faults in the 9th round of AES key scheduling was enough to break the encryption system. For active attacks, the most common approach is that of fault injections, as detailed in [44].

Countermeasures to fault injections include established techniques in communications engineering, such as the use of error codes and parity checking, along with newer proposals such as concurrent error detection (CED) which suppress the operation of a circuit when error states are detected. The aim of CED is to halt the propagation of the error to the output, where the attacker can analyse whether the fault attack was successful or not. Additional proposals for countermeasures include the duplication of circuitry, or repeated computation, to provide comparators. With duplication of hardware the cost penalty is high, while with repeated computation the execution time may increase significantly. Other, more efficient, schemes have been proposed, such as suggested in [45], requiring only one parity bit for each internal state of AES. The approach detects all odd errors, and in many cases the even errors, and may be a promising approach for implementation in both the hardware and software contexts.

Proposals have also been made to secure the CRT computations of RSA. In [46], the arguments of the CRT were calculated using an approach termed efficient redundancy, where values are verified before their use in the RSA algorithm. This approach, which adds little timing overhead, improves upon previous approaches requiring full redundancy.

3.2.9 Out-of-order execution attacks

At the time of writing, two new side-channel attacks [47], targeting the out-of-order execution of instructions on processors, were announced. Meltdown exploits the scenario where a speculatively executed instruction, although aborted, permits the bypassing of memory protections and thus the ability to read Kernel memory from user space. The attack is deemed to affect Intel processors primarily. In the short-term, a patch based on the KAISER countermeasure of [48] has been released. This countermeasure re-maps the memory space in software. A more permanent solution will likely require architectural changes at the hardware level to control the order of permission checks for access to memory and improvements to memory segmentation.

The Spectre attack exploits the use of speculative branch predictions to store information to cache memory that can then be targeted with side-channel techniques such as flush+reload or evict+reload cache attacks. The attack is considered more universal than Meltdown, and has already been shown to affect Intel, AMD and ARM processors. Countermeasures against Spectre also appear difficult to implement. Simply disabling speculative execution would result in an unacceptable performance loss, while inserting temporary blocking instructions is also seen as a challenging task. Potential updates to processor microcode may be possible as a form of software patch, but likely to impact performance considerably.

4. Conclusions

The move from cloud deployment model to the edge has implications for security. In contrast to a cloud data centre, housed within a large building complex with a significant level of security, the edge deployment will constitute a large number of small clusters or individual installations, where high levels of physical security are not economically viable. In many situations, physical security of the micro-server may consist primarily of a light-weight enclosure, designed to protect the system from environmental factors and vandalism or casual tampering efforts. For the determined attacker, this may not prove to be an effective barrier and it should be assumed that a realistic worst-case scenario is that an attacker will be able to gain full access to the system. This then creates a larger threat surface, now incorporating physical attacks that can be used to compromise the individual micro-server, and potentially, the wider network.

Deployment at the edge still requires the implementation of traditional server and network security practises, such as those outlined in this chapter. In addition, deployment at the edge should assume that networks are operating over untrustworthy links and therefore the use of encrypted tunnelling through VPNs, and the use of malware detection, firewalls, intrusion detection/prevention systems and DNSSEC should all be considered as forming the basis of an endpoint security policy.

The use of virtualisation is a core element of cloud and resource sharing technologies; however, it also opens the possibility for attacks exploiting VMEscape. Accommodating guests with differing security levels, such as DMZ and internal, on the same host, should be avoided.

Edge deployment should consider the further threats posed from an attacker gaining partial, or full, physical access to a system. This requires input not only from a hardware security standpoint, but also from software perspectives. Applications developers should employ secure coding practises, particularly when operating on any sensitive information, as highlighted in the discussions of memory attacks in Section 2.2.1. Care should also be taken to minimise or, if possible, to avoid the storage of secret information in physical memory, since attacks such as buffer overflows and removal of frozen DRAM modules have been shown as effective means to extract information stored in the clear. User passwords, for example, should be stored as hashed values and passwords requested on demand for comparison or verification. The use of software, or ideally hardware-based, hard disk encryption technologies can offer protections, even when the disk is removed from a system.

Side-channel attacks can potentially be used to reveal sensitive information such as the extended margin information stored in the log and policy files. Indeed, the variation of voltage and frequency margins, core features of the UniServer solution, may also influence the relative amount of side-channel leakages. A countermeasure to this threat is the deployment of encryption using side-channel resilient countermeasures, such as masking, to break the statistical link between power measurements and hypothetical power models.

In the full stack deployment, representing a micro-server data centre, the UniServer software is running under the host OS, abstracted from guest applications operating under VMs. However, in the bare metal deployment, the UniServer software runs along-side other system applications. It is in this deployment architecture where the UniServer system is most exposed to interference by other applications, which can potentially view and access each other's files or resources. The UniServer log files were identified as high value assets that need to be protected from tampering, since it could potentially lead to system instability or denial of service attacks. It is therefore a recommendation that the log and policy files are stored in an encrypted format, to avoid reading and manipulation by others. Additionally, consideration should be given as to whether the files should be digitally signed, to provide assurance that they come from a trusted source. These recommendations would naturally have overheads in terms of real-time operation, so their implementation would need to be considered carefully in terms of system performance. The use of encryption, and possibly digital signing, will likely be candidate to form a security solution.

Author details

Charles J. Gillan and George Karakonstantis*
The School of Electrical and Electronic Engineering and Computer Science (EEECS), Queen's University Belfast, Northern Ireland

*Address all correspondence to: g.karakonstantis@qub.ac.uk

References

[1] Stojmenovic I, Wen S. The fog computing paradigm: Scenarios and security issues. In: 2014 Federated Conference on Computer Science and Information Systems (FedCSIS). Warsaw, Poland: IEEE; 2014. pp. 1-8

[2] Alrawais A, Alhothaily A, Hu C, Cheng X. Fog computing for the internet of things: Security and privacy issues. IEEE Internet Computing. 2017;**21**(2):34-42

[3] Mukherjee M, Matam R, Shu L, Maglaras L, Ferrag MA, Choudhury N, et al. Security and privacy in fog computing: Challenges. IEEE Access. 2017;**5**:19293-19304

[4] Cabric M. Corporate Security Management. Challenges, Risks, and Strategies. 1st edition. London: Butterworth-Heinemann; 2015:242. ISBN: 9780128029343

[5] S Institute. Operating System Security and Secure Operating Systems [Online]. Available from: https://www.giac.org/paper/gsec/2776/operating-system-security-secure-operating-systems/104723 [Accessed: December 2017]

[6] IBM. Business Intelligence Architecture and Deployment Guide—Securing the Operating System [Online]. Available from: https://www.ibm.com/support/knowledgecenter/en/SSEP7J_10.2.1/com.ibm.swg.ba.cognos.crn_arch.10.2.1.doc/c_securing_the_operating_system.html [Accessed: December 2017]

[7] Wang Z, Jiang X, Cui W, Ning P. Countering kernel rootkits with lightweight hook protection. In: Proceedings of the 16th ACM Conference on Computer and Communications Security. 2009. pp. 545-554

[8] Friedl S. An Illustrated Guide to the Kaminsky DNS Vulnerability [Online]. Available from: http://unixwiz.net/techtips/iguide-kaminsky-dns-vuln.html [Accessed: December 2017]

[9] Kocher P. Timing attacks on implementations of Diffie-Hellman, RSA, DSS, and other systems. In: Advances in Cryptology (CRYPTO '96). Lecture Notes in Computer Science. Vol. 1109. Heidelberg, Berlin: Springer; 1996. pp. 104-113

[10] Aly H, ElGayyar M. Attacking AES using Bernstein's attack on modern processors. In: Youssef A, Nitaj A, Hassanien AE, editors. Progress in Cryptology – AFRICACRYPT 2013. Vol. 7918. Lecture Notes in Computer Science. Berlin, Heidelberg: Springer; 2013. pp. 127-139

[11] Page D. Theoretical use of cache memory as a cryptanalytic side-channel. IACR Cryptology ePrint Archive. 2002:1-23

[12] Tsunoo Y, Saito T, Suzaki T, Shigeri M, Miyauci H. Cryptanalysis of DES implemented on computers with cache. Cryptographic Hardware and Embedded Systems-CHES. 2003;**2003**

[13] Tromer E, Osvik D, Shamir A. Efficient cache attacks on AES, and countermeasures. Journal of Cryptology. 2010;**23**(1):37-71

[14] Kocher P, Jaffe J, Jun B. Differential power analysis method and apparatus. U.S. Patent 7587044; 2009

[15] Cryptocoding.net. Cryptographic Coding Standards. Available from: https://cryptocoding.net/index.php/Cryptography_Coding_Standard; 2013

[16] NaCl: Networking and Cryptography library. Available from: https://nacl.cr.yp.to/

[17] Tsunoo Y, Saito T, Suzaki T, Shigeri M, Miyauchi H. Cryptanalysis

of DES implemented on computers with cache. In: Walter CD, Koç ÇK, Paar C, editors. Cryptographic Hardware and Embedded Systems - CHES 2003. Vol. 2779. Lecture Notes in Computer Science. Berlin, Heidelberg: Springer; 2003. pp. 62-76

[18] Tromer E, Osvik DA, Shamir A. Efficient cache attacks on AES, and countermeasures. Journal of Cryptology. 2010;**23**:37-71. DOI: 10.1007/s00145-009-9049-y

[19] Kocher P. Timing attacks on implementations of Diffie-Hellman, RSA, DSS and other systems. In: Advances in Cryptology - CRYPTO'96. Berlin: Springer; 1996. pp. 104-113

[20] Qiao R, Seaborn M. A new approach for rowhammer attacks. In: 2016 IEEE international symposium on Hardware oriented security and trust (HOST). McLean, Virginia, USA: IEEE; 2016

[21] Brier E, Clavier C, Olivier F. Correlation Power Analysis with a Leakage Model in Cryptographic Hardware and Embedded Systems - CHES. In: Joye M, editor. Berlin, Heidelerg: Springer; 2004:16-29. DOI: 10.1007/978-3-540-28632-5_2

[22] Mangard S, Oswald E, Popp T. Power Analysis Attacks: Revealing the Secrets of Smart Cards. New York, USA: Springer-Verlag US; 2007

[23] Tiri K, Akmal M, Verbauwhede I. A dynamic and differential CMOS logic with signal independent power consumption to withstand differential power analysis on smart cards. In: Proceedings of the 28th European Solid-State Circuits Conference (ESSCIRC 2002). University of Bologna; 2002

[24] Quisquater JJ, Samyde D. Electromagnetic analysis (EMA): Measures and counter-measures for smart cards. In: Smart Card Programming and Security (E-smart 2001). Lecture Notes in Computer Science. Berlin, Heidelberg: Springer; Vol. 2140. 2001. pp. 200-210

[25] Agrawal D, Archambeault B, Rao JR, Roha P. The EM side-channel(s). In: Cryptographic Hardware and Embedded Systems (CHES). Lecture Notes in Computer Science. Berlin, Heidelberg: Springer; Vol. 2523. 2002. pp. 29-45

[26] Yamaguchi M, Kobayashi S, Sugawa T, Toriduka H, Homma N, Satoh A, et al. Development of an on-chip micro shielded-loop probe to evaluate performance of magnetic film to protect a cryptographic LSI from electromagnetic analysis. In: International Symposium on Electromagnetic Compatibility (EMC). Lauderdale , Florida: IEEE; 2010. pp. 103-108

[27] Fahn P, Pearson P. IPA: A new class of power attacks. In: Cryptographic Hardware and Embedded Systems. Lecture Notes in Computer Science. Berlin, Heidelberg: Springer; Vol. 1717. 1999. pp. 173-186

[28] Tiri K, Verbauwhede I. Charge recycling sense amplifier based logic: Securing low power security ICs against differential power analysis. In: Proceedings of the 30th European Solid-State Circuits Conference, ESSCIRC. IEEE; Sept 2004. pp. 179-182

[29] Hospodar G, Gierlichs B, De Mulder E, Verbauwhede I, Vandewalle J. Machine learning in side-channel analysis: A first study. Journal of Cryptographic Engineering. 2011;**1**(4):293-302

[30] Heuser A, Zohner M. Intelligent machine homicide: Breaking cryptographic devices using support vector machines. In: Constructive Side-Channel Analysis and Secure Design—COSADE 2012. Vol. 7275. Series LNCS. Berlin, Heidelberg: Springer; 2012. pp. 249-264

[31] Lerman L, Bontempi G, Markowitch O. Side channel attack: An approach based on machine learning. In: Constructive Side-Channel Analysis and Secure Design—COSADE. 2011

[32] Markowitch O, Medeiros S, Bontempi G, Lerman L. A machine learning approach against a masked AES. Journal of Cryptographic Engineering. 2013;**5**(2):62-75. DOI: 10.1007/s13389-014-0089-3

[33] Gilmore R, Hanley N, O'Neill M. Neural network based attack on a masked implementation of AES. In: IEEE International Symposium on Hardware Orientated Security and Trust. IEEE; 2005

[34] Anderson R, Kuhn M. Tamper resistance—A cautionary note. In: Proceedings of Second USENIX Workshop on Electronic Commerce. Oakland, California: USENIX; 1996. pp. 1-11. Available from: https://www.usenix.org/legacy/publications/library/proceedings/ec96/index.html

[35] Anderson R, Kuhn M. Low cost attacks on tamper resistant devices. In: Proceedings of 5th Security Protocols Workshop. Lecture Notes in Computer Science. Berlin, Heidelbergs: Springer; Vol. 1361. 1997. pp. 125-136

[36] Boneh D, DeMillo R, Lipton R. On the importance of eliminating errors in cryptographic computations. Journal of Cryptology. 2001;**14**(2):101-119

[37] Piret G, Quisquater J-J. A differential fault attack technique against SPN structures, with application to the AES and Khazad. In: Walter CD, Koç CK, Paar C editors. Proceedings of the 5th International Workshop, Cologne, Germany, September 8-10. Vol. 2779. Lecture Notes in Computer Science. Berlin, Heidelberg: Springer; 2003. pp. 77-88. ISBN: 978-3-540-45238-6

[38] Kim CH, Quisquater J-J. New differential fault analysis on AES key schedule: Two faults are enough. In: Grimaud G, Standaert FX, editors. Smart Card Research and Advanced Applications. Lecture Notes in Computer Science. Vol. 5189. Berlin, Heidelberg: Springer; 2008. pp. 48-60. DOI: 10.1007/978-3-540-85893-5_4

[39] Bar-Eli H, Choukri H, Naccache D, Tunstall M, Whelan C. The sorcerer's apprentice guide to fault attacks. Proceedings of the IEEE. 2006;**94**(2):370-382

[40] Bertoni G, Breveglieri L, Koren I, Maistri P, Piuri V. Error analysis and detection procedures for a hardware implementation of the advanced encryption standard. IEEE Transactions on Computers. 2003;**52**(4):492-505

[41] Shamir A. Method and apparatus for protecting public key schemes from timing and fault attacks. U.S. Patent 5991415; 1999

[42] The Real Story of Stuxnet. IEEE Spectrum Online [Online]. 2013. Available from: https://spectrum.ieee.org/telecom/security/the-real-story-of-stuxnet

[43] USBKiller V3. USBKill [Online]. Available from: https://usbkill.com/ [Accessed: December 2017]

[44] Lipp M, Schwarz M, Gruss D, Prescher T, Haas W, Mangard S, et al. Meltdown and Spectre. Bugs in Modern Computers Leak Passwords and Sensitive Data [Online]. 2018. Available from: https://meltdownattack.com/meltdown.pdf

[45] Gruss D, Lipp M, Schwarz M, Fellner R, Maurice C, Mangard S. KASLR is dead: Long live KASLR. In: International Symposium on Engineering Secure Software and Systems. Austria: University of Graz; 2017

[46] Kocher P, Genkin D, Gruss D, Haas W, Hamburg M, Lipp M. Meltdown and Spectre. Bugs in Modern Computers Leak Passwords and Sensitive Sata [Online]. 2018. Available from: https://spectreattack.com/spectre.pdf

[47] Watson RNM, Woodruff J, Roe M, Moore SW, Neumann PG. Capability Hardware Enhanced RISC Instructions (CHERI): Notes on the Meltdown and Spectre Attacks. University of Cambridge Technical Reports, UCAM-CL-TR-916, Feb 2018. Available from: https://www.cl.cam.ac.uk/techreports/UCAM-CL-TR-916.pdf

[48] Gruss D, Lipp M, Schwarz M, Fellner R, Maurice C, Mangard S. KASLR is dead: Long live KASLR. In: Bodden E, Payer M, Athanasopoulos E, editors. Engineering Secure Software and Systems. ESSoS 2017. Lecture Notes in Computer Science. Vol. 10379. Cham: Springer; 2017. pp. 161-176

Chapter 5

Securing the Deployment of Cloud-Hosted Services for Guaranteeing Multitenancy Isolation

Laud Charles Ochei

Abstract

Multitenancy introduces significant error and security challenges in the cloud depending on the location of the functionality to be shared and the required degree of isolation between the tenants. Existing approaches for securing the deployment of cloud-hosted services to serve multiple users have paid little attention to evaluating the effect of the varying degrees of multitenancy isolation on the security and access privilege of tenants (or components). In addition, approaches for securing the isolation of tenants (or components) are usually implemented at lower layers of the cloud stack and often apply to the entire system and not to individual tenants (or components). This study presents CLAMP (Cloud-based architectural approach for securing services through Multitenancy deployment Patterns) to securing the deployment of cloud-hosted services in a way that guarantees the required degree of isolation between the tenants. We evaluated the framework by applying it to a motivating cloud deployment problem. The findings show among other things that the framework can be used to select suitable deployment patterns, evaluate the effect of varying degrees of isolation on the cloud-hosted service, analyse the deployment requirements of cloud-hosted services and optimise the deployment of the cloud-hosted service to guarantee multitenancy isolation.

Keywords: security, cloud-hosted services, deployment, multitenancy, tenant isolation

1. Introduction

Applications on the cloud are accessed over the internet using standard internet protocols. In deciding to store data or host applications in the public cloud, an organisation loses its ability to access the servers that store its information. In this way, potentially sensitive data are at risk from insider attacks.

Therefore, cloud service providers must put in place security measures to physical access to the servers in the data center and frequently monitor data centers for suspicious activity. Security and privacy challenges deriving from the use of the internet are substantial and but no different from the security issues of the applications not hosted in the cloud. The one significant security element introduced by the cloud is multitenancy [1].

Multitenancy is an essential cloud computing property. Multitenancy is a software architecture where one instance of a cloud offering is used to serve multiple tenants and/or components [2, 3]. Multitenancy means that your application is utilising a virtual machine on a physical computer that is hosting multiple virtual machines. There are many forms of attack utilising multitenancy- inadvertent data sharing, virtual machine escape, side channel attack, and denial of service attack.

Users can require varying or different degrees of isolation between components when implementing multitenancy. To avoid interference, a high degree of insulation between components may be required, but this usually results in high resource consumption and running costs per component. A low degree of isolation promotes sharing of components, resulting in low resource consumption and running costs, but with high performance impact when the workload changes and the application does not scale up/down.

The challenge therefore is how to: (i) ensure that there is isolation between multiple tenants accessing the service or components designed (or integrated) with the service; (ii) resolve the trade-offs between varying degrees of isolation between tenants or components.

Motivated by this problem, this study presents a framework, CLAMP (Cloud-based architectural approach for securing services through Multitenancy deployment Patterns) to securing the deployment of cloud-hosted services in a way that guarantees the isolation between tenants. The framework assumes that the issues of security are tackled from the perspective of the tenant owns software components and is responsible for configuring them to design and deploy its own cloud-hosted application on a shared cloud platform whose provider does not have control over these components.

We evaluated the framework by applying it to a motivating cloud deployment problem that requires securing several components of a cloud-hosted service while guaranteeing the required degree of isolation between tenants. The research question addressed in this study is: "How can we secure the deployment of cloud-hosted services in a way that guarantees isolation between tenants".

The main contributions of this study are:

1. To develop a framework for securing the deployment of cloud-hosted services in a way that guarantees the isolation of tenants.

2. To evaluate the framework by applying it to a motivating cloud deployment problem.

3. To develop a cloud security checklist for guiding software architects in implementing the framework.

4. Present recommendations and best practice guidelines for securing the deployment of cloud-hosted services based on the framework.

Our findings show among other things that the framework can be used to select suitable deployment patterns, evaluate the effect of varying degrees of isolation on the cloud-hosted service, analyse the deployment requirements of cloud-hosted services and optimise the deployment of the cloud-hosted service to guarantee multitenancy isolation.

The rest of this chapter is organised as follows. Section 2 presents an overview of cloud computing and cloud security. Section 3 presents architectures for cloud-hosted services. Section 4 presents multitenancy in a cloud environment. Section 5 discusses related work on multitenancy and cloud security. Section 6 presents a framework for securing the deployment of cloud-hosted services for guaranteeing multitenant isolation, while Section 7 evaluates the framework by applying it to a motivating cloud deployment problem. Section 8 provides further discussion and recommendations for securing the deployment of cloud-hosted services based on the framework. Section 9 concludes the chapters with future work.

2. Cloud computing security

This section gives an overview of cloud computing and cloud security and multitenancy.

2.1 Cloud computing

According to Armbrust et al. [4], "cloud computing refers to both the applications delivered as a service over the Internet and the hardware and systems software in the data centers that provides those.
services."

The cloud includes hardware for the data centre as well as software. The cloud could either be a *public cloud* (that is, cloud that is provided to the general public in a prepaid manner), *private cloud* (that is, an organisation's internal IT infrastructure which is not available to the public at large), or a *hybrid cloud* (that is, a private cloud's computing capacity that is enhanced by the public cloud).

Although there are so many definitions that have been given for the term cloud computing, there is common agreement on the basic characteristics of a cloud computing environment. These include [3]—pay-per-use, elastic capacity and the illusion of infinite, self-service interface, and resources that are abstracted or virtualized.

There are three basic cloud service models:

i. *Software as a Service (SaaS):* In the SaaS model, cloud providers can install, operate and access their application software using a web browser. An example of a SaaS provider is Salesforce.com, which utilises the SaaS model to provide Customer Relationship Management (CRM) applications located on their server to customers. This eliminates the need for customers to run and install the application on their own computers.

ii. *Platform as a Service (PaaS):* In the PaaS model, cloud providers deliver cloud platforms which represent an environment for application developers to create and deploy their applications. A notable example of PaaS is the Google App Engine, which provides an environment for creating and deploying web-based applications written in specific programming languages.

iii. *Infrastructure as a Service (IaaS):* In the IaaS model, cloud providers offer physical (computers, storage) and virtualized computer resources. Examples of IaaS providers include: Amazon EC2, and Azure Services Platform.

2.2 Cloud security

Cloud security relates to a wide range of policies, techniques, applications, and controls used to safeguard virtualized IP, information, apps, services, and related infrastructure. Cloud security is very essential for companies making the shift to the cloud and also for customers who use the cloud for a range of personal services especially as security threats continue to evolve and become more advanced. Cloud security concerns fall into two wide classifications: (i) security concerns faced by cloud providers (businesses providing software, platform, or infrastructure-as - a-service organisations through the cloud); (ii) security concerns faced by their customers (businesses or organisations that host applications or store data in the cloud). However, the responsibility is shared. There are four (4) main forms of attack that use multitenancy: inadvertent information sharing, virtual machine escape, side-channel attack, denial of service attack. The focus of this study is mostly related to inadvertent information sharing where a tenant has a set of components/resources or services which are mapped to some physical resource on the cloud platform. Under this situation, data residing on the physical resource from one tenant may be leak to another tenant.

Cloud service suppliers often store more than one customer information on the same server in order to conserve resources (e.g., CPU, memory, storage space) reduce cost and maintain service level agreement. To handle such sensitive situations, cloud service providers usually put in place robust secure measures to ensure proper data isolation and logical storage segregation [5].

Cloud security is the protection of data, applications, and infrastructures involved in cloud computing. Cloud security concerns can be grouped in various ways. Gartner listed seven (7) categories of cloud security. In the "data segregation" category, which is the closest to the focus of our study, the cloud is typically in a shared environment alongside data from other customers [6]. The Cloud Security Alliance identified 12 areas of concern [7]. In "Abuse and Nefarious Use of Cloud Services" category, which is the closet to our study, the focus is on the use of poorly secured cloud service deployments, free cloud service trials and fraudulent account sign-ups via payment instrument fraud expose cloud computing models such as IaaS, PaaS, and SaaS to malicious attacks.

3. Architectures for cloud-hosted services

The architectures or cloud patterns used to deploy cloud-hosted services to the cloud are of great importance to software architects because they determine whether or not the system's essential quality attributes (e.g., performance) will be exhibited [1, 8, 9].

3.1 Architectural patterns

Architectural and design patterns have long been used to provide known solutions to many common problems a distributed system face [1, 10]. A system/application architecture decides whether or not it will show its necessary quality attributes (e.g., performance, availability, and security) [1, 8].

Definition 2.3: **Architectural Pattern.** Architectural patterns are compositions of architectural elements that provide bundled solutions to solve recurring problems a system faces [1].

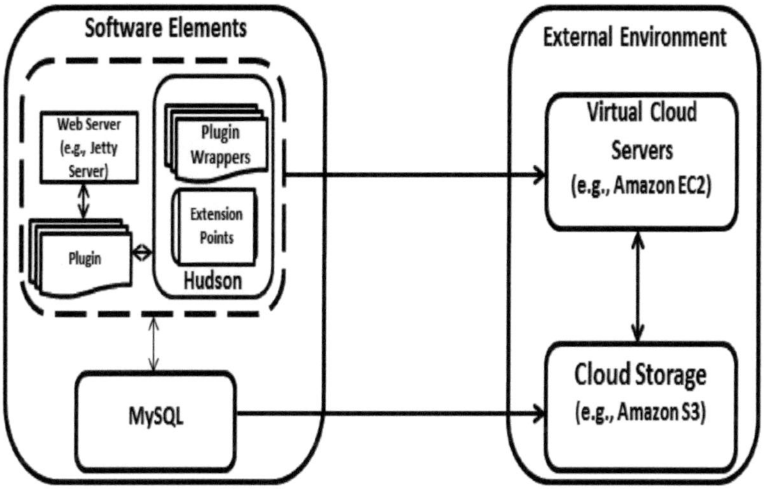

Figure 1.
Mapping elements of a cloud-hosted service to the external environment.

A cloud pattern in the cloud computing environment represents a well-defined format for explaining an appropriate solution to a cloud-related problem [11]. There are several cloud problems, such as: (i) selecting an appropriate cloud type for hosting applications; (ii) selecting a cloud service delivery approach; (iii) deploying a multi-tenant service in a way that ensures tenant isolation.

Cloud deployment architects are using cloud patterns as a reference guide to document best practice on how to plan, develop and deploy cloud-based applications.

Definition 2.4: **Cloud Deployment Pattern.** A "Cloud deployment pattern" is defined as a type of architectural pattern, which embodies decisions as to how elements of the cloud application will be assigned to the cloud environment where the application is executed.

Our definition of cloud deployment pattern is similar to the concept of design patterns [10], (architectural) deployment patterns [1], collaboration architectures [8], cloud computing patterns [11], cloud architecture patterns [12], and cloud design patterns [13].

One of a cloud deployment architect's main duty is to assign cloud application elements to the hardware elements (e.g. processor, filesystems) and communication elements (e.g. protocols, message queues) in the cloud environment so that the necessary quality attributes can be achieved.

Figure 1 demonstrates how elements of Hudson (a typical of Global Software Development tool) are mapped to the elements of the cloud environment. Hudson operates on an Amazon EC2 instance while periodically extracts and stores the data it produces on separate cloud storage (e.g., Amazon S3).

4. Multitenancy in a cloud environment

Multitenancy is an essential cloud computing property where a single instance of a cloud offering is used to serve multiple tenants and/or components [14, 15]. One of the challenges of implementing multitenancy on the cloud is how to enable the required degree of isolation between multiple components of a cloud-hosted application (or tenants accessing a cloud-hosted application). We refer to this as *multitenancy isolation*.

Definition 1: **Multitenancy isolation.** The term "Multitenancy Isolation" refers to an approach to ensuring that one tenant's performance, stored data volume, and access rights do not impact other tenants accessing the shared application component or its functionality. Multitenancy isolation can be represented in three main cloud multitenancy patterns [11]:

1. Shared component: Tenants use the same instance of a resource and may not be aware that other tenants are using it.

2. Tenant-isolated component: Tenants share the same resource instance but are assured of their isolation. This pattern allows for the tenant-specific configuration of the functionality or resource offered.

3. Dedicated component: Tenants do not share resource instance. That is, each tenant is associated with one instance (or a certain number of instances) of the resource.

4.1 Degrees of multitenancy isolation

The degree of isolation between tenants accessing a shared component of an application can be expressed in the three multitenancy patterns (i.e., shared component, tenant-isolated component and dedicated component). The shared component reflects the lowest degree of isolation between tenants whilst the highest is the dedicated component.

The three key areas where tenant isolation can be addressed in a system are: performance, stored data volume and access privileges. For example, in performance isolation, other tenants should not be affected by the workload created by other tenants. For example, other tenants should not be impacted by the workload generated by other tenants when considering performance isolation.

Guo et al. [16] evaluated different isolation capabilities related to authentication, information protection, faults, administration etc.

Different isolation capabilities related to faults, information protection, authentication, administration, etc., have been evaluated by Guo et al. [16]. Bauer and Adams [17] have studied how to virtualization can be used to ensure that the failure of one tenant instance does not spread into other tenant instances.

A high degree of isolation can be achieved by deploying an application component exclusively for one tenant. This would ensure that there is little or no performance interference between the components when workload changes. The deployment of an application component specifically for one tenant can achieve a high degree of insulation. This ensures that when workload changes, there is little or no performance impact between the components.

Nevertheless, since components are not shared (e.g. in a situation where some strict laws and regulations prohibit them from being shared), this means duplicating the components for each tenant, resulting in high resource consumption and running costs. In general, this would restrict the number of requests to access the components.

It may also be that a component requires a low degree of isolation, for example, to facilitate sharing of the functionality, data, and resources of the component. This would minimise resource consumption and running costs, but other component's performance might be affected if one of the components experiences a change in workload.

The challenge for a cloud deployment architect would therefore be how to overcome the trade-offs between the required performance, system resources and access privileges at different levels of an application when selecting one (or combinations) of multitenancy patterns to deploy software tools in the cloud. Resolving the trade-off involving access privileges of users at different levels of an application depending on the type of multitenancy deployment pattern that is being used is one of the strategies for providing security for cloud-hosted services deployed based on multitenancy architecture.

4.2 Implementation of multitenancy isolation

Multitenancy isolation can be implemented both at the process levels (i.e., based on the processes that interacts with the system) and data levels (i.e., based data that is being generated or manipulated by the system) of a cloud-hosted service. **Figure 2** shows an architecture that can be used to implement multitenancy isolation at the data level. This implementation represents an application that logs each operation into a database by relying on an automated build verification and testing in response to a specific event such as detecting changes in a file.

A specific example of an implementation shown in **Figure 2** is to use Hudson's Files Found-Trigger plugin to poll one or more directories and start a build if there are certain files in those directories [18]. Hudson is an open source tool and so can be easily modified by adding a Java class that accepts a filename as argument into the plugin. The plugin is loaded into a separate class loader during execution, to avoid interfering with the core functionality of Hudson.

Definition 2: **Application Component.** This refers to an encapsulation of a functionality or resource that is shared between multiple tenants. A component of an application could be a data handling component (e.g. database), communication

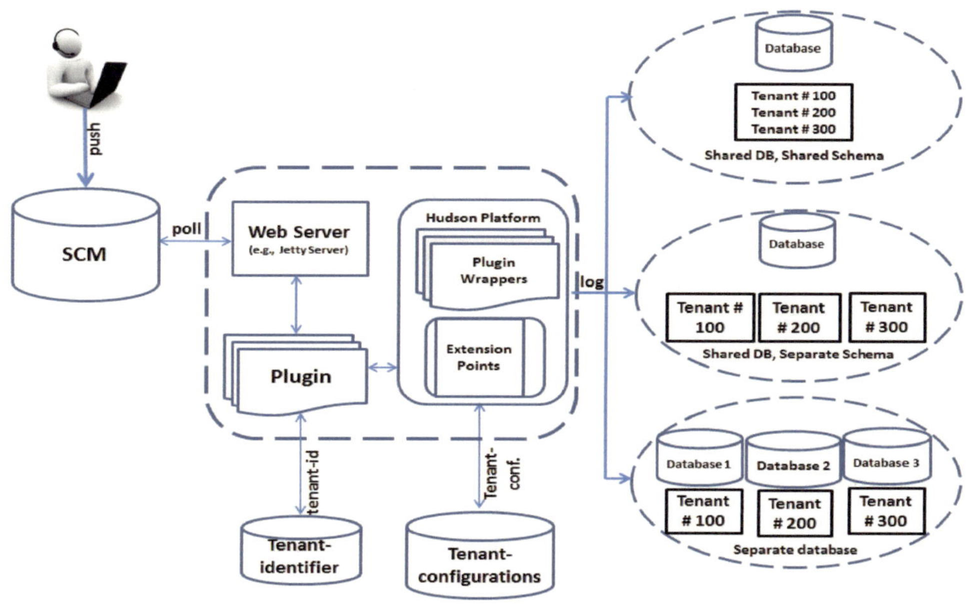

Figure 2.
Multitenancy isolation architecture for cloud-hosted applications.

component (e.g. message queue), user interface component (e.g. AJAX) or processing component (e.g. load balancer).

There are several solutions to multitenancy implementation which have been widely discussed in the literature. Multitenancy can be introduced at different cloud stack layers: application layer [16], middleware layer [19], and data layer [20, 21].

It has been suggested that customization is the solution to addressing the hidden constraints on multitenancy such as complexities, security, scalability and flexibility [22]. Furthermore, integrating a plugin into a cloud-based service can provide a workaround for true multitenancy. Again, most of the solutions available to incorporate multitenancy require a re-engineering of the cloud service to some degree [17, 23].

Other research work on multitenancy isolation include: [24–30].

5. Related work on cloud security

Apart from the general research on best practices in securing the cloud against various forms of attacks, there is little research on approaches to secure cloud services against attacks arising from implementing multitenancy architectures. There is also little research on approaches for securing the deployment of cloud-hosted services in a way that guarantees varying degrees of isolation between tenants.

According to Bass et al., one of the significant security challenges introduced in the cloud is multitenancy [1]. Implementing multitenancy means that your cloud-hosted services are utilising the virtual machine on a physical machine that host multiple virtual machines. Much of literature on multitenancy and cloud security has established that the obvious approach to addressing the problem is for cloud providers to allow users to reserve entire virtual machines for their use. Although this defeats some of the economic benefits of using the cloud, it is nevertheless a mechanism to prevent multitenancy attacks [1–3].

Previous research has looked at this problem from the perspective of the cloud providers, for instance, autoscaling algorithms and supporting security-based strategies provided by IaaS providers such as Amazon and optimization frameworks suggested for use by SaaS providers such as Salesforce.com.

This study, however, looks at the issue from the tenant's viewpoint, who owns software components and is responsible for configuring them to build and deploy their own cloud-hosted application on a shared cloud platform where the cloud provider has no control over the software components. The focus of this chapter is to provide a framework for securing the deployment of cloud-hosted services in a way that guarantees multitenancy isolation.

The work by [31] is one of the most detailed studies on cloud security. The author explores different aspects of security and the possible solutions that have been considered by different authors. The author did not consider approaches for securing the deployment of cloud-hosted services in a way that guarantees varying degrees of isolation between tenants.

6. Framework for securing the deployment of cloud-hosted services for guaranteeing multitenant isolation

The section discusses the framework for securing the deployment of cloud-hosted services for guaranteeing multitenant isolation.

6.1 Developing the CLAMP framework

The study presents a robust framework, CLAMP, for securing the deployment of cloud-hosted services for guaranteeing multitenancy isolation. The framework, CLAMP (Cloud-based architectural approach for securing services through Multitenancy deployment Patterns), is basically a framework for guiding software architects in securing the deployment of cloud-hosted services in a way that guarantees the required degree of isolation between other tenants when one of the tenants (or components) experiences a high workload or security breach.

The CLAMP framework is illustrated as a layered architecture in **Figure 3**. It shows how the components of the framework work together to support the task of securing the deployment of components of a cloud-hosted service for guaranteeing multitenancy isolation. The development of CLAMP was inspired by the well understood architectural structure/pattern called layered pattern [1]. A layer is an abstract "virtual machine" that provides a cohesive set services through a managed interface. In a strictly layered system, a layer can only use the services of the layer immediately below it. This structure is used to imbue a system with portability, the ability to change the underlying computing platform.

The different components of the CLAMP framework are described as follows.

6.1.1 Layer one: selection of a suitable architectural pattern

This layer addresses the selection of a suitable architectural pattern. In order to secure the deployment of cloud-hosted services for guaranteeing multitenancy isolation, it may be very difficult if not impossible to use one cloud pattern to deploy the service to the cloud due to the different requirements of the service including accessibility of the service to a wider audience and a combined assurance for security and privacy. For instance, the architect would require a combination of

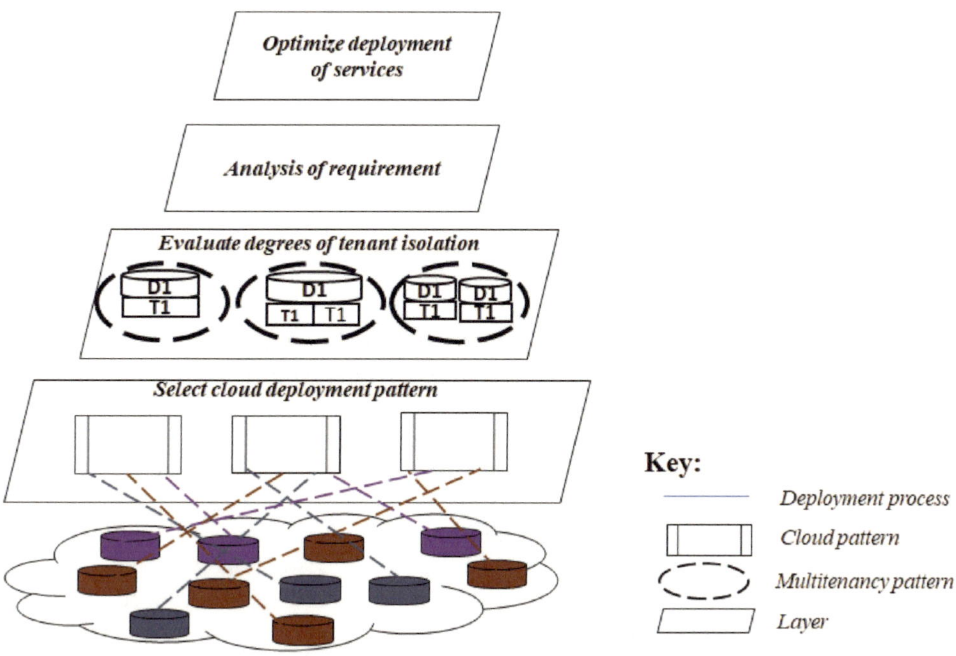

Figure 3.
A layered architecture for securing the deployment of cloud-services for guaranteeing multitenancy isolation.

several deployment patterns together with supporting technologies for archiving components of the cloud-hosted service (i.e., in a hybrid fashion) to integrate components located in a different cloud environment to form one cloud solution. Again, if communication is required internally to exchange messages between application components, then a message-oriented middleware technology would also be needed. Therefore, the challenge is that of selecting a suitable pattern (together with the supporting technologies) or a combination of patterns in order to secure the deployment of cloud-hosted services for guaranteeing multitenancy isolation. It is assumed that there is a repository of cloud deployment patterns from where a software architect can select a suitable pattern (s) to address the business requirements of the company/user.

6.1.2 Layer two: evaluation of the required degree of isolation between tenants

The layer addresses the evaluation of the required degree of isolation between tenants. There are varying degrees of isolation between tenants that are accessing a cloud-hosted service. Some of the tenants would require a higher or different degree of isolation than others. Tenants would be able to share application components as much as possible at the very basic degree of multitenancy, which translates into increases use of underlying resources.

At the very basic degree of multitenancy, tenants would be able to share application components as much as possible which translates to increased utilisation of underlying resources. While some components of the application may benefit from a low degree of isolation between tenants, other components may require a higher degree of isolation because the component may be either too sensitive or cannot be shared as a result of certain corporate legislation and regulation.

There is increasing evidence, for example, that many cloud providers are reluctant to set up data centres in mainland Europe due to stricter legal requirements that prohibit data processing outside Europe [32, 33]. This requirement will traverse down to the IaaS level, and customers must take this into consideration if intending to host applications outsourced to such cloud providers [11] that host customers data outside Europe. Therefore, evaluating the required degree of isolation between the tenants will allow for the appropriate mapping of security requirements during the deployment of cloud-hosted services onto cloud provider's infrastructure.

6.1.3 Layer three: analysis of the deployment requirements of the cloud-hosted service

Layer three addresses the analysis of the deployment requirements of the cloud-hosted service. This involves two main activities: (i) mapping tenant isolation to key process of the cloud-hosted services, cloud resources required to support the service and layers of the cloud stack on the which the service will be executed; (ii) analysing the trade-offs that should be considered when implementing the required degree of tenant isolation.

The mapping is rooted in the framework of a typical architectural deployment system that has two main components: the cloud application (that is, the component or service to be deployed) and the cloud environment (that is, the environment in which the process/service is performed) [1]. This mapping also captures the link between a process associated with a cloud-hosted service (e.g., continuous integration process), being used in a hybrid deployment scenario by utilising a cloud-hosted environment (e.g., SaaS and PaaS deployment environment).

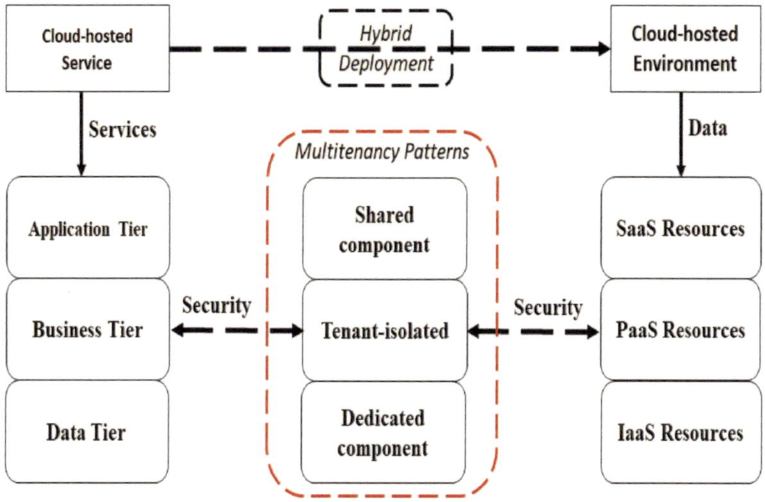

Figure 4.
Mapping of degrees of tenant isolation to cloud-hosted services and resources.

In our previous research, we provided an explanatory framework for (i) mapping tenant isolation to different software processes, cloud resources and application stack layers (ii) illustrating the different trade-offs for consideration in order to achieve optimal deployment of components in a way that guarantees the required degree of tenant isolation [34] (see **Figure 4**).

Issues relating to security, privacy, trust and regulatory compliance can mostly be tackled in a hybrid fashion. For example, data / bugs created from a bug tracking system could be stored at some location to comply with privacy and legal regulations, while the architecture of the bug tracking system could be changed to limit the access of certain data to users residing in regions not deemed to be of interest to those who own the hosted data. Securing cloud-hosted services deployed with the goal of guaranteeing varying degrees of multitenancy isolation can best be tackled using a hybrid approach.

The second aspect of the analysis involves analysing the key trade-offs for consideration when implementing the required degree of tenant isolation for cloud-hosted software processes. There are six key aspects of the trade-offs that have to be considered when implementing security for multi-tenant cloud-hosted software services. These trade-off include tenant isolation versus (resource sharing, the number of users/requests, customizability, the size of generated data, the scope of control of the cloud application stack and business constraints). **Table 1** shows the trade-offs and the key decision that have to be main when considering the trade-offs.

6.1.4 Layer four: optimisation of the deployment of the cloud-hosted services

This layer deals with the optimization of the components of a cloud-hosted service. In a cloud environment, varying degrees of tenant isolation are possible, depending on the type of component being shared, the process supported by the component and the location of the component on the cloud application stack (i.e., application level, platform level, or infrastructure level).

In a cloud environment, depending on the type of component being shared, the processes enabled by the component, and the location of the component on the cloud application stack (i.e. application level, platform level, or network level),

Category	Checklist
Selection of a suitable architectural pattern	What are classes of cloud patterns available, what are the tools and processes to support the selection of suitable cloud patterns.
Evaluation of the required degree of isolation between tenants	What are the data and processes of the cloud-hosted service that require security? What is the required degree of isolation between tenants accessing the components of the cloud-hosted services?
Analysis of the deployment requirements of the cloud-hosted	How can you map the key resources of the cloud-service (e.g., store for the archive data) to the cloud provider's platform? What are the trade-offs to consider when securing the deployment of cloud-hosted services? (e.g., customizability, scope of control, business requirements)
Optimisation of the deployment of the cloud-hosted services	What are the components (or tenants) that are required to design (or integrate) with the cloud-hosted services? How feasible is it to tag components or whole system?

Table 1.
Security checklist for evaluating the framework.

varying degrees of tenant isolation are possible. Therefore, it is important for software architects to be able to able to control the required degree of isolation between tenants sharing components of a cloud-hosted application.

For instance, the deployment of an application component specifically for one tenant will achieve a high degree of isolation. This would make sure that when workload changes, there is little or no performance impact between the components.

However, because components are not shared it implies duplicating the components for each tenant, which leads to high resource consumption and running cost. Overall, this will limit the number of requests allowed to access the components. A low degree of isolation would allow sharing of the component's functionality, data and resources. This would reduce resource consumption and running cost, but the performance of other components may be affected when one of experiences a change in workload.

This is a decision-making challenge that requires an appropriate decision to be made to address the trade-off between a lower degree of isolation versus the possible influence that can occur between components or a high degree of isolation versus the difficulty of high resource usage and component running costs.

In a nutshell, the procedure for implementing the framework can be summaries with following four steps: (i) Select suitable deployment patterns (one or combination of several patterns), (ii) Evaluate the effect of varying degrees of isolation on the cloud-hosted service, (iii) Analyse the deployment requirements of cloud-hosted services and (iv) optimise the deployment of the cloud-hosted service to guarantee multitenancy isolation.

6.2 Developing a security checklist for deployment of cloud-hosted services

In addition to the framework, CLAMP, we develop a security checklist to guide software architects in securing the deployment of cloud hosted services. The layers of the frameworks are used to develop the categories of the checklist. Many of the items in the checklist may seem obvious but the purpose of a checklist is help ensure the completeness of the security design while implementing the CLAMP framework.

In using the security checklist, the software architect should think about how to review the security of the cloud-hosted services and figure out how well it satisfies security in each of the categories of the framework. In other words, what questions would you ask a software architect to evaluate how the framework satisfies the requirements for securing the deployment of cloud-hosted services for guaranteeing multitenancy isolation. This is the basis for the security checklist.

7. Evaluation of framework for securing the deployment of cloud-hosted services

This section presents a simple case study of a cloud deployment problem to illustrate how to use the proposed framework to secure the deployment of a cloud-hosted services in a way that guarantees multitenancy isolation. The following scenario explains our motivation.

7.1 Motivating scenario

Let us assume that there are multiple components of a cloud service (e.g., data-handling component) hosted on the same or different cloud infrastructure. These components which are of various types and sizes are required to design (or integrate with) a cloud-hosted service (e.g., continuous integration system such as Hudson or Jenkins) and their supporting processes for deployment to multiple tenants. Tenants, in this case, may be multiple users, departments of a company or different companies. The laws and regulations of the company make it liable to share and archive data generated from the component (e.g., builds of source code) and keep it accessible for auditing purposes. However, access to some components or some aspects of the archived data will be provided solely to particular groups of tenants for security reasons. The question is: in a resource-constrained environment, how can we secure the deployment of components of this cloud-hosted service in a way that guarantees the required degree of isolation between other tenants when one of the tenants (or components) experiences a high workload or security breach (**Table 2**).

7.2 Applying the CLAMP framework

This section explains how to apply the proposed framework, CLAMP, to secure the deployment of this cloud-hosted service in a way that guarantees the required degree of isolation between other tenants. Each component of the framework has

Category	Analysis
Selection of a suitable architectural pattern	The problem requires a hybrid-related deployment pattern, namely, integrating data stored in multiple clouds
Evaluation of the required degree of Isolation between tenants	The requirement to allow a particular group of users to access some components for security reasons means that the company requires the highest degree of isolation between tenants
Analysis of the deployment requirements of the cloud-hosted	Map the tenant isolation to key processes associated with the cloud-hosted service, cloud resources and layers of the cloud stack. Analyse the trade-offs required for optimal deployment
Optimisation of the deployment of the cloud-hosted services	Tag each component. Analyse the trade-off involved, namely, achieving a high degree of isolation versus resource sharing. To address this trade-off, an optimization model is recommended to be used to select optimal components for deployment to the cloud

Table 2.
Summary of how problem was analysed per layer of the framework.

a part to play in securing the deployment of components of a cloud-hosted service. The structure of evaluating the framework, CLAMP, in its textual form, is specified as a string consisting of three sections-(i) Context; (ii) Problem; and (iii) Solution. In a more general sense, the string can be represented as: [CONTEXT; PROBLEM; SOLUTION]. Each layer of the framework maps to the step required to provide a solution to the cloud deployment problem. **Table 2** summaries how the problem was evaluated based each layer of the framework.

7.2.1 Step one: selecting a suitable cloud deployment pattern

In order to address this challenge, this framework would recommend that the architect should reference some sort of a classification or taxonomy to guide in the selection of a suitable pattern together with the supporting technologies. In our previous work, we have developed a taxonomy and a process for guiding architect in selecting a suitable framework for cloud deployment [35]. In addition, a general process, CLIP (CLoud-based Identification process for deployment Patterns) has been developed for guiding architects in selecting applicable cloud deployment patterns (together with the supporting technologies) using the taxonomy for deploying services/application to the cloud we also discussed.

It is important to note that the company does not have direct access to the cloud IaaS. Therefore, the architect must select a deployment pattern that can be implemented at the application level to secure the deployment of the cloud-hosted services for guaranteeing multitenancy isolation. By making reference to the taxonomy of cloud-deployment patterns and the general process for selecting applicable deployment patterns based on the taxonomy, we would recommend that the architect should select a hybrid-related deployment pattern for addressing the requirements of the customer. It is assumed that the data archived by Hudson contains the source code and (possibly configuration files) that drives a critical function of an application used by the company.

The data stored by Hudson is presumed to contain the source code and (possibly configuration files) which drives a critical function of an application used by the company. Any unauthorised access to it may be devastating for the company. In this circumstance, the most appropriate multitenancy pattern to use is the hybrid backup deployment pattern. This pattern can be used to extract data to the cloud environment and archive it different cloud environments [11].

7.2.2 Step two: evaluating the varying degrees of isolation

This step involves evaluating the required degree of isolation between tenants and then select an appropriate multitenancy pattern or combination of patterns to support such a required degree of isolation. There are varying degrees of isolation between tenants that are accessing the cloud-hosted service and so some of the tenants would require a higher or different degree of isolation than others.

One of the key requirements of the company to provide access to some components or some aspects of the archived data solely to particular groups of tenants for security reasons. Based on this key requirement, we conclude that the company requires the highest degree of isolation between tenants.

The varying degrees of multitenancy isolation can be captured in three main cloud deployment patterns: shared component, tenant-isolated component and dedicated component. The shared component represents the lowest degree of isolation between tenants while the dedicated component represents the highest. In a dedicated component pattern, tenants do not share resources, though each tenant is associated with one instance or a certain number of instances of the resource.

7.2.3 Step three: analysis of the deployment requirements of the cloud-hosted service

The step involves analysing the deploying requirements of the cloud-hosted services. This analysis entails mapping tenant isolation to key processes associated with the cloud-hosted service, cloud resources required to support the service and layers of the cloud stack on the which the service will be executed. This analysis translates to using a hybrid approach to map the SaaS and PaaS level of the cloud provider to the cloud-hosted service which has a backup cloud storage. This type of cloud pattern is referred to as a hybrid backup pattern [3]. The archive data in a problem scenario can be stored in any location to comply with privacy and legal regulations of the company while the architecture of the cloud-hosted service could be modified to restrict exposure of certain data to users located in regions not considered to be of interest to the owners of the hosted data.

The second aspect of the analysis involves analysing the different trade-offs to be considered for optimal deployment of components with a guarantee of the required degree of tenant isolation. There are three main trade-offs that the company has to consider. The first trade-off relates to tenant isolation versus customizability. The higher the degree of isolation that is required, the easier it is to customise a cloud-hosted service to implement tenant isolation. However, because we assumed that the user has access to the application layer of the cloud stack, it would be more difficult to implement a higher degree of isolation at the application level in terms effort, time and skills set required to modify the source code. This raises issues of compatibility and interdependencies between the cloud-hosted services and required plugins and libraries. Each time a multitenant application or its deployment environment changes, then a tedious, complex and security maintenance process is also required.

The second trade-off relates to the “scope of control” of the cloud application stack. The architect has more flexibility to implement or support the implementation of the required degree of tenant isolation when there is greater “scope of control” of the cloud stack application. As the company requires a higher degree of isolation (e.g., based on the dedicated component), then the scope of control should extend beyond the higher level to the lower levels of the cloud stack (i.e., PaaS and IaaS) even as the cost of implementation of such a cloud security architecture will certainly increase.

The third trade-off relates to the trade-off between tenant isolation and business (or legal) requirements of the company. A key legal requirement of the company is that access to some components or some aspects of the archived data will be provided solely to particular groups of tenants for security reasons. The dedicated component which offers a high degree of isolation can be used to handle the legal requirements Such legal restriction, for example, legal restrictions and the location and configuration of the cloud infrastructure are usually difficult to compensate for at the application level. For example, a legal requirement can state that data that a specific cloud provider has hosted in Europe cannot be stored elsewhere (e.g., in the USA). Therefore an architect would have to map this form of requirement to a cloud infrastructure that specifically meets this requirement.

7.2.4 Step four: optimisation of the deployment of the cloud-hosted services

The key task in step four is to optimise the deployment of components of the cloud-hosted service. Some requirements cannot be fully satisfied and so there has to be some optimisation to ensure that the cloud deployment is carried out in way that does not compromise the security of the components of the cloud-hosted service. This entails tagging the components (or tenants) associated with the cloud-hosted service so that the software architects can be have more leverage to

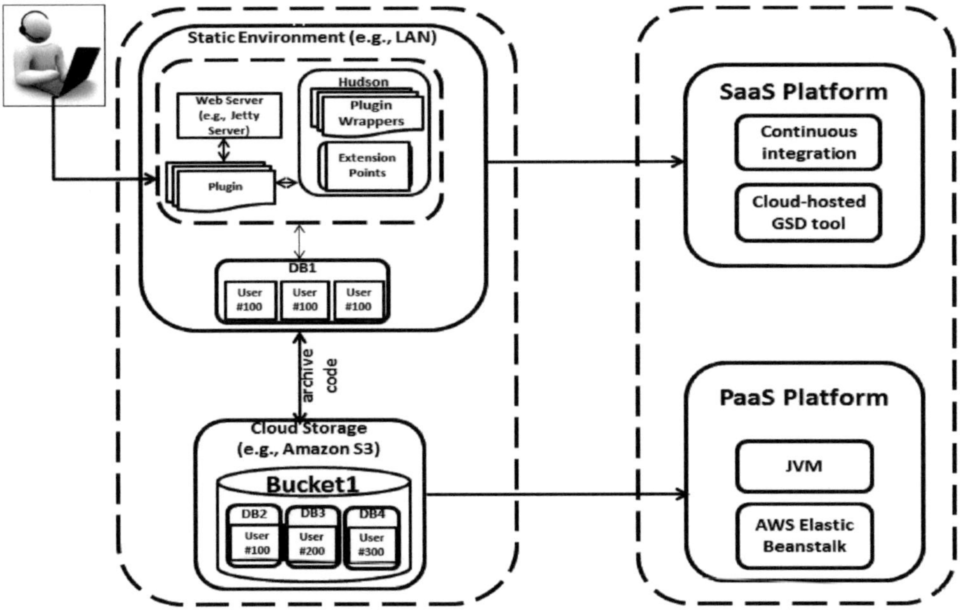

Figure 5.
Mapping a continuous integration system to cloud stack based on hybrid backup pattern.

implement the required degree of isolation between tenants. In [36] an implementation of the model-based algorithm was presented for providing optimal solutions for deploying components designed to use (or be integrated with) a cloud-hosted application in a way that guarantees multitenancy isolation (**Figure 5**).

7.3 Applying the security checklist

In addition to applying the framework on the motivating problem, we also apply the security checklist to support design and analysis of process for securing the deployment of cloud-hosted services for guaranteeing multitenancy isolation. **Table 3** shows the result of the security checklist.

Category	Checklist
Selection of a suitable architectural pattern	The hybrid patterns are a class of cloud pattern that can be explored. The hybrid backup pattern is suitable for the problem. Tools and technologies such as cloud storage, and REST, and message exchange technologies can be implemented
Evaluation of the required degree of isolation between tenants	The highest degree of isolation would be required for isolate tenants. The data to secure include archive data- source code, configuration files. The key software process in this problem is the continuous integration process
Analysis of the Deployment requirements of the Cloud-hosted	The process supporting the cloud-hosted service (i.e., continuous integration) should be mapped to a cloud platform that allows data to be stored in multiple location without much restrictions. The key trade-offs in this problem are tenant isolation versus (customizability, scope of control, business requirements)
Optimisation of the deployment of the cloud-hosted services	The main components to optimise are—authorization/authentication data or database components, queue messages. The approach of tagging components can be done either manually or dynamically using a model/algorithm depending on the number of components and complexity of the processes involved

Table 3.
Applying the security checklist.

8. Discussions and recommendations

This section presents a general discussion of the key security issues that should be considered together with some recommendations that can be followed in order to secure the deployment of cloud-hosted services in a way that guarantees multitenancy isolation.

8.1 Assurance for compliance with legislation and regulatory requirements

One of the challenges of implementing cloud security is to provide assurance to cloud users who need to demonstrate compliance with various legislation and regulatory requirements. Our proposed framework addresses this challenge by providing guidance to the software architecture based on the a taxonomy of cloud deployment patterns to not only to select a suitable cloud deployment pattern but to also evaluate the requirements of the customer to select a cloud multitenancy pattern that guarantees the required degree of isolation between tenants.

For example, there is growing evidence that many cloud providers are unwilling to set data centres in mainland Europe because of tighter legal requirements that disallow the processing of data outside Europe (Hon & Millard 2017, Google 2017). This requirement will traverse down to the IaaS level, and customers must take this into consideration if intending to host applications outsourced to such cloud providers [11]. The challenge, therefore, for a cloud deployment architect is that there are no case studies to understand and evaluate the effect of the required degree of isolation on the performance, systems resources and access privileges at different levels of a cloud-hosted service when opting for one (or combinations) of a particular degree of isolation between tenants.

8.2 Customizability of the cloud-hosted services and supporting process

Customising a cloud-hosted GSD tool (or any cloud-hosted service) can be very challenging if the service has several components that are being shared. A service deployed on the cloud can have many inter-dependencies on different levels of the application itself and with other applications, plugins, libraries, etc., deployed with other cloud providers. This could impact the security of the cloud-hosted system in a way that we did not anticipate and thus the degree of tenant isolation that was needed. There is also a serious risk that incompatible plugins and libraries will be used to alter, configure and run these GSD tools. This could corrupt the GSD tool and stop other supporting programs/processes from running. A simple way to tackle this infrastructure problem is to move tenant isolation deployment down the lower levels of the cloud stack, where the architect can deploy the GSD framework on a PaaS platform, for example. Middleware issues and methods for SaaS device customizability were discussed in [37, 38].

8.3 Errors and sensitivity to workload interference

Multitenancy may pose significant error and security challenges in the cloud, particularly when different degrees of isolation are introduced between multiple tenants who share resources. When resources are shared between multiple tenants in a multitenant cloud-service, it is very possible to affect the performance and resource usage of other tenants due to errors associated with one tenant (e.g. due to overload of the tenant or inadequate resource allocated to the tenant).

The type of error associated with a cloud-hosted service is a pointer to the key resources to consider in achieving the required degree of tenant isolation.

For example, moving the VM image instance associated with a cloud hosted service whose file permission had been set on a local machine to the cloud infrastructure could cause affect the requires degree of tenant isolation and hence the security of other tenants during cloud deployment. Therefore, it is necessary to get repository ownership and permission right before deploying such a cloud-hosted service.

8.4 Tagging components with the required degree of isolation

One of the challenges of securing the deployment of a cloud-hosted service is how to handle such cloud-hosted services that several interdependencies with other services elements to which it interacts. Therefore, it is important that components designed to be used or incorporated with a cloud-hosted service should be tagged as much as possible when the necessary degree of tenant isolation is needed.

Tagging can be a complex and complicated process and may not even be feasible under certain circumstances (e.g. where the component is incorporated into other systems and is not under customer control). Therefore, this can also be predicted in a dynamic way instead of labelling each part with an insulation value as necessary.

In our previous work [39], we built an algorithm that dynamically learns the features of existing components in a repository and then uses this knowledge to associate each component with the appropriate degree of isolation. This information is critical to making key security decisions and optimising the resources consumed by the components, particularly in a dynamic or real-time environment.

9. Concluding remarks

The chapter presented CLAMP, a framework for securing the deployment of cloud-hosted services in a way that guarantees the isolation between tenants to contribute to the literature on multitenancy and cloud security. The framework is based on a layered architectural structure where the layers are allowed to use other layers in a strictly managed fashion; a layer is only allowed to use the layer immediately below.

The framework was evaluated by applying it to a motivating cloud deployment problem that requires securing several components of a cloud-hosted service while guaranteeing the required degree of isolation between tenants. The findings show among other things that the framework can be used to select suitable deployment patterns, evaluate the effect of varying degrees of isolation on the cloud-hosted service based on the requirements of the business, analyse the deployment requirements of cloud-hosted services and optimise the deployment of the cloud-hosted service to guarantee multitenancy isolation.

Future work would entail design an experimental procedure for automatically evaluating the framework (i.e., the layered-architectural structure) for securing the deployment of a real-life cloud-hosted service for guaranteeing isolation between tenants. Thereafter, this experimental design will incorporate into a simulator and testing tool for evaluating the layered-architecture for securing the cloud-hosted service for guaranteeing isolation between tenants. This approach has been discussed in [1] as a way to turn architectural parameters into constants, ranges and other that can be easily measured. This will allow software architects to determine the effect of each form of improvement or business requirements of the component or cloud-hosted service before deciding whether the service is secured enough to be deployed without compromising the required degree of isolation between tenants.

Author details

Laud Charles Ochei
Robert Gordon University, Aberdeen, United Kingdom

*Address all correspondence to: l.c.ochei@rgu.ac.uk

References

[1] Bass L, Clements P, Kazman R. Software Architecture in Practice, 3/E. United States: Pearson Education; 2013

[2] Bauer E, Adams R. Reliability and Availability of Cloud Computing. New Jersey: John Wiley & Sons; 2012

[3] Buyya R, Broberg J, Goscinski A. Cloud Computing: Principles and Paradigms. New Jersey, United States: John Wiley & Sons, Inc.; 2011. DOI: 10.1002/9780470940105

[4] Armbrust M, Fox A, Griffith R, Joseph AD, Katz R, Konwinski A, et al. A view of cloud computing. Communications of the ACM. 2010;**53**(4):50-58

[5] Srinivasan MK, Sarukesi K, Rodrigues P, Manoj MS, Revathy P. State-of-the-art cloud computing security taxonomies—A classification of security challenges in the present cloud computing environment. In: Proceedings of the International Conference on Advances in Computing, Communications and Informatics; ACM. 2012. pp. 470-476

[6] Brodkin J. Gartner—Seven Cloud-Computing Security Risks. 2019. Available from: https://www.infoworld.com/article/2652198/gartner–seven-cloudcomputing-security-risks.html [Accessed: 14 August 2019]

[7] Brook J-M, Field S, Shackleford D. Top threats to cloud computing plus: industry insights. 2019. Available from: https://cloudsecurityalliance.org/artifacts/top-threats-cloud-computing-plusindustry-insights/ [Accessed: 14 August 2019]

[8] Junuzovic S, Dewan P. Response times in n-user replicated, centralized, and proximity-based hybrid collaboration architectures. In: Proceedings of the 2006 20th Anniversary Conference on Computer Supported Cooperative Work; ACM. 2006. pp. 129-138

[9] Stol K-J, Avgeriou P, Babar MA. Design and evaluation of a process for identifying architecture patterns in open source software. In: Ivica C, Volker G, Matthias B, editor. Software Architecture: 5th European Conference, ECSA 2011, Essen, Germany, September 13-16, 2011. Proceedings. Vol. 6903. London: Springer; 2011. pp. 147-163

[10] Vlissides J, Helm R, Johnson R, Gamma E. Design Patterns: Elements of Reusable Object-Oriented Software. Vol. 49. Boston, United States: Addison-Wesley; 1995. p. 120

[11] Fehling C, Leymann F, Retter R, Schupeck W, Arbitter P. Cloud Computing Patterns. London, England: Springer; 2014

[12] Wilder B. Cloud Architecture Patterns. 1st ed. Sebastopol, CA, United States: O'Reilly Media, Inc.; 2012

[13] Homer A, Sharp J, Brader L, Narumoto M, Swanson T. Cloud Design Patterns. Redmon, Washington, United States: Microsoft; 2014

[14] Krebs R, Momm C, Kounev S. Metrics and techniques for quantifying performance isolation in cloud environments. Science of Computer Programming. 2014;**90**:116-134

[15] Pearson S. Privacy, security and trust in cloud computing. In: Privacy and Security for Cloud Computing. London: Springer-Verlag; 2013. pp. 3-42

[16] Mehta A. Multi-tenancy for cloud architectures: Benefits and challenges. 2017. Available from: http://www.devx.

com/architect/Article/47798/ [Accessed: May 2020]

[17] Aiken L. Why multi-tenancy is key to successful and sustainable software-as-a-service (SaaS). 2017. Available from: http://www.cloudbook.net/resources/stories/. [Accessed: May 2020]

[18] Hudson. Apache Software Foundation. 2016. Available from: http://wiki.hudson-ci.org//display/HUDSON/Files+Found+Trigger [Accessed: May 2020]

[19] Strauch S, Andrikopoulos V, Leymann F, Muhler D. Esb mt: Enabling multi-tenancy in enterprise service buses. In: In 2012 IEEE 4th International Conference on Cloud Computing Technology and Science (CloudCom); IEEE. 2012. pp. 456-463

[20] Vengurlekar N. Isolation in private database clouds. 2012. Available from: http://www.oracle.com/technetwork/database/database-cloud/ [Accessed: May 2020]

[21] Wang ZH, Guo CJ, Gao B, Sun W, Zhang Z, An WH. A study and performance evaluation of the multi-tenant data tier design patterns for service oriented computing. In: IEEE International Conference on E-Business Engineering, 2008. ICEBE'08; IEEE. 2008. pp. 94-101

[22] Khan MF, Mirza AU, et al. An approach towards customized multi-tenancy. International Journal of Modern Education and Computer Science. 2012;**4**(9):39

[23] Momm C, Krebs R. A qualitative discussion of different approaches for implementing multi-tenant saas offerings. In: Software Engineering (Workshops). Vol. 11. 2011. pp. 139-150

[24] Mietzner R, Unger T, Titze R, Leymann F. Combining different multitenancy patterns in service-oriented applications. In: Enterprise Distributed Object Computing Conference, 2009. EDOC'09; IEEE International; IEEE. 2009. pp. 131-140

[25] Yusoh ZIM, Tang M. Composite saas placement and resource optimization in cloud computing using evolutionary algorithms. In: 2012 IEEE 5th International Conference on Cloud Computing (CLOUD); IEEE. 2012. pp. 590-597

[26] Shaikh F, Patil D. Multi-tenant e-commerce based on saas model to minimize it cost. In: In 2014 International Conference on Advances in Engineering and Technology Research (ICAETR); IEEE. 2014. pp. 1-4

[27] Westermann D, Momm C. Using software performance curves for dependable and cost-efficient service hosting. In: Proceedings of the 2nd International Workshop on the Quality of Service-Oriented Software Systems; ACM. 2010. p. 3

[28] Abbott ML, Fisher MT. The Art of Scalability: Scalable Web Architecture, Processes, and Organizations for the Modern Enterprise. Indiana, United States: Pearson Education; 2009

[29] Leymann F, Fehling C, Mietzner R, Nowak A, Dustdar S. Moving applications to the cloud: An approach based on application model enrichment. International Journal of Cooperative Information Systems. 2011;**20**(03):307-356

[30] Aldhalaan A, Menasc, e DA. Near-optimal allocation of VMS from iaas providers by saas providers. In: 2015 International Conference on Cloud and Autonomic Computing (ICCAC); IEEE. 2015. pp. 228-231

[31] Singh A, Chatterjee K. Cloud security issues and challenges: A survey. Journal of Network and Computer Applications. 2017;**79**:88-115

[32] Hon K, Millard C. Eu data protection law and the cloud. International Association of Privacy Professionals; 2020. Available from: https://iapp.org/resources/article/ [Accessed: February 2020]

[33] Google. Google cloud platform and the eu data protection directive. Google Inc.; 2020. Available from: https:// cloud.google.com/security/compliance/ eu-data-protection/ [Accessed: February 2020]

[34] Ochei LC, Bass J, Petrovski A. Degrees of tenant isolation for cloud-hosted software services: A cross-case analysis. Journal of Cloud Computing: Advances, Systems and Applications. 2018;**7**(22)

[35] Ochei LC, Bass JM, Petrovski A. A novel taxonomy of deployment patterns for cloud-hosted applications: A case study of global software development (gsd) tools and processes. International Journal on Advances in Software. 2015;**8**(3-4):420-434

[36] Ochei LC, Petrovski A, Bass JM. Optimal deployment of components of cloud-hosted application for guaranteeing multitenancy isolation. Journal of Cloud Computing. 2019;**8**(1):1

[37] Walraven S. Middleware and methods for customizable SaaS [PhD thesis]. KU Leuven: Department of Computer Science; 2014, 2014. p. 6

[38] Walraven S, Van Landuyt D, Truyen E, Handekyn K, Joosen W. Efficient customization of multi-tenant software-as-a-service applications with service lines. Journal of Systems and Software. 2014;**91**:48-62

[39] Ochei LC, Petrovski A, Bass J. An approach for achieving the required degree of multitenancy isolation for components of a cloud-hosted application. In: 4th International IBM Cloud Academy Conference (ICACON 2016). 2016

Chapter 6

Semantic Web and Interactive Knowledge Graphs as an Educational Technology

Victor Telnov and Yuri Korovin

Abstract

Technologies of knowledge representation, inductive reasoning, and semantic annotation methods are considered in relation to knowledge graphs that are focused on the domain of nuclear physics and nuclear power engineering. Interactive visual navigation and inductive reasoning in knowledge graphs are performed using special search widgets and an intelligent RDF browser. As a toolkit for ontologies refinement and enrichment, a software agent for the context-sensitive searching for new knowledge in the WWW is presented. In order to evaluate the measure of compliance of the found content with respect to a specific domain, the binary Pareto relation and Levenshtein metrics are used. The proposed semantic annotation methods allow the knowledge engineer to calculate the measure of the proximity of an arbitrary network resource in relation to classes and objects of specific knowledge graphs. Operations with remote semantic repositories are implemented on cloud platforms using SPARQL queries and RESTful services. The proposed software solutions are based on cloud computing using DBaaS and PaaS service models to ensure scalability of data warehouses and network services. Examples of using the proposed technologies and software are given.

Keywords: knowledge database, ontology, inductive reasoning, knowledge graph, semantic annotation, cloud computing, education

1. Introduction

Nowadays, the ontology description languages RDF, OWL [1], description logics [2], and knowledge graphs provide a modern theoretical basis for the creation of systems and methods of acquisition, presentation, processing, and integration of knowledge in computer systems of artificial intelligence.

There are substantial considerations in favor of the predominant use of inductive reasoning in modern knowledge graphs instead of traditional deduction. Inductive reasoning rules based on consideration of possible alternatives (precedents) allow generating and verifying cognitive hypotheses (fuzzy knowledge) that cannot be obtained directly by deductive reasoning in the graph. Inductive inference is one of the basic technologies of semantic annotation of the WWW content, when it is necessary to refine, expand, and update existing graphs with new knowledge. With the help of the inductive inference, the problems of classification and clustering of new entities in the semantic database of nuclear knowledge are solved [3].

The aim of the work presented in the chapter is to create a working prototype first and then a semantic web portal of knowledge in the domain of nuclear physics and nuclear power engineering based on ontologies and using databases deployed on cloud platforms [3]. The task of the study was to create the following graphs of nuclear knowledge:

- World nuclear data centers
- Nuclear research centers
- Events and publications from CERN
- IAEA databases and network services
- Nuclear physics at MSU and MEPhI
- Nuclear physics journals
- Integrated nuclear knowledge graph

To ensure the effective use of the nuclear knowledge database in educational activities, additional software agents have been created for reconnaissance context-sensitive search for adequate network content and its semantic annotation based on existing knowledge graphs (for example, with the aim of authoring training materials), as well as public endpoints for easy navigation on international knowledge databases DBpedia and Wikidata.

The potential beneficiaries of information solutions and technologies that are proposed in the chapter are students, professors, experts, engineers, managers, and specialists in the domain of nuclear physics and nuclear power engineering (target audience).

2. Knowledge representation: ontology design

Ontologies are often regarded as special knowledge repositories that can be read and understood both by people and computers, alienated from the developer and reused. Ontology in the context of information technology is a formal specification with a hierarchical structure, which is designed to represent knowledge. Typically, ontology includes descriptions of classes of entities (concepts) and their properties (roles) in relation to a certain subject domain of knowledge, as well as relationships between entities and restrictions on how these relationships can be used. Ontologies, which additionally include objects (instances of entity classes) and particular statements about these objects, are also called knowledge graphs. The formal ontology model O is understood as an ordered triple of the form

$$\mathrm{O} = <X, R, F>, \text{ where}$$

X is a finite set of entity classes (concepts) for the domain represented by the ontology O; R is a finite set of properties (roles) that establish relationships between entities for some domain; and F is a finite set of interpretation functions defined on entities and/or properties for ontology O. It can be said that interpretation functions map formal ontologies to certain domains.

As an illustration of the ontology creation process, **Figure 1** below shows a design pattern for an ontology "Nuclear Training Center" type, which is used in the project [4]. This model was created on the basis of an analysis of the educational programs of the following Russian and international training centers: National Research Nuclear University MEPhI, Physics Department of Moscow State University, IAEA. The ontology design pattern is represented in the UML notation according to the international standard [5]. The actual ontology in serialized form for the knowledge graph titled "Nuclear Physics at MSU and MEPhI" is available in Ref. [6]. Another approach to the development and refinement of the structure of ontologies is based on Terminological Decision Trees (TDT) [7].

One of the attractive features of the semantic web is that it becomes possible to extract (infer) new knowledge from the facts which already exist in the knowledge graph. For this purpose, intelligent software agents are used, which are called reasoners. The way inference is carried out algorithmically is not specified in the ontology itself or in the corresponding OWL document, since OWL is a declarative language for ontologies describing. The correct answer to any question is determined by the semantics of the description logic that sets the language standard.

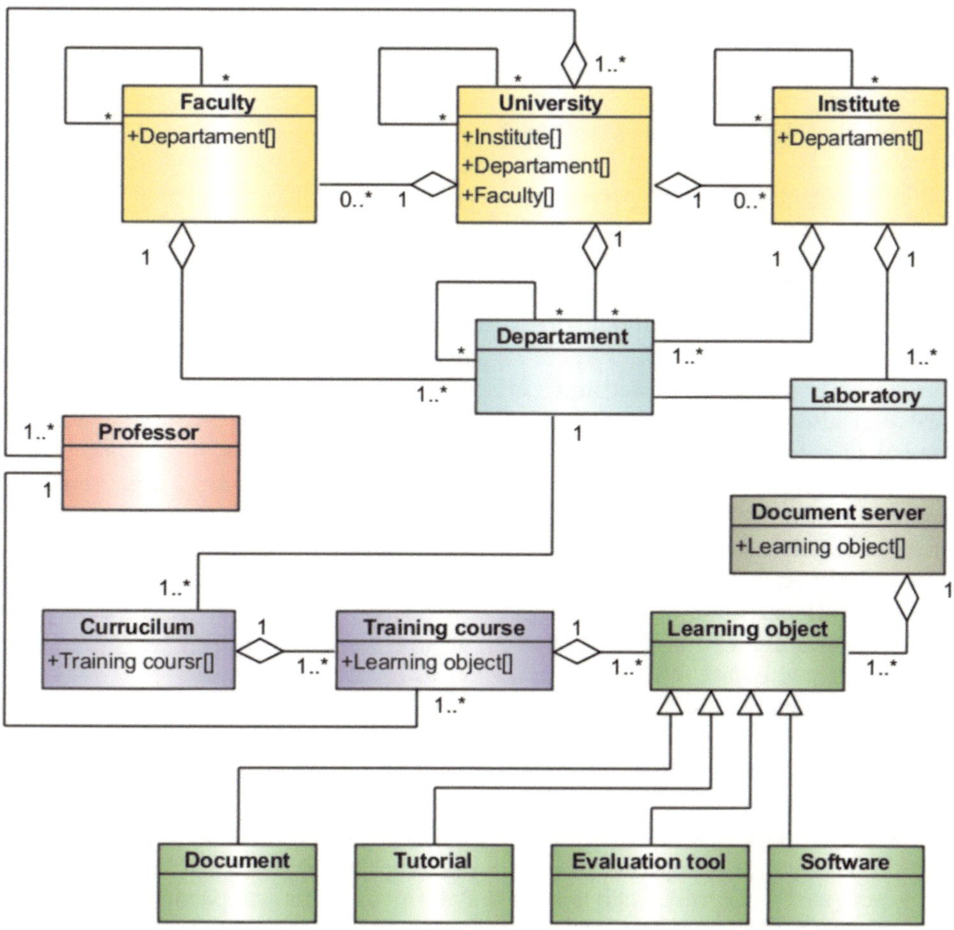

Figure 1.
Design pattern for an ontology "nuclear training center" type in UML notation.

In particular, the project under discussion is based on the description logic with the signature SROIQ (D), see [3].

The RDF browser is another significant attribute of the project [4], which distinguishes it from other well-known solutions in the field of semantic web. An example of inductive reasoning using the RDF browser is given below. Clusters of entities that are related to each other by a particular property or group of properties are examples of deduced facts (samples of new knowledge) that were not originally explicitly presented in the graph. The deduced facts in the RDF browser have the form of petals grouped around the nodes of the graph, are opened with a mouse click, and are very convenient for subsequent visual navigation in the graph.

Once on the desired location of the desired knowledge graph using the search widget, then the user through the RDF browser can perform visual navigation on the graph, visiting its nodes in the desired order and extracting metadata, hypertext links, full-text, and media content associated with the node, wherein the neighborhood (environment, closure) of each node of the graph becomes visible and available for navigation. This neighborhood includes the nodes of the graph, through which the user initially entered the semantic web, as well as adjacent nodes of other graphs that are supported by the knowledge database [3].

The visual way of specifying the inference rules on the graph makes it stand out from the more traditional reasoner's interfaces, where inference rules are specified using SWRL language, logical predicates or a SPARQL-like syntax. It seems, that the intuitive, interactive visual way of specifying inference rules is more friendly for unsophisticated users of knowledge graphs.

3. Inductive reasoning in knowledge graphs

Knowledge graphs may contain various kinds of uncertainties. For this reason, the presentation of real domains of knowledge in the context of the semantic web may encounter difficulties if only classical logical formalisms are used. Alternative approaches sometimes assume the probabilistic nature of knowledge, which is hardly always appropriate and justified [8]. In addition, purely deductive exact logical reasoning may not be possible for knowledge databases on the WWW; such reasonings do not take into account statistical patterns in the data. In this regard, of particular interest is the ability of knowledge databases as artificial intelligence systems to evaluate cognitive hypotheses, using for this purpose, in addition to a deduction, other methods of reasoning, such as inductive reasoning, argumentation, and reasoning based on precedents.

As an example of inductive reasoning in the knowledge graph, consider the following situation [3]. Some students have to pass an exam in nuclear physics at the Physics Department of Moscow State University. Let the student know only the title of the training course: "Physics of the atomic nucleus and particles" and the name of the professor: "I.M. Kapitonov." Let us formulate the task.

Task 1. Using the semantic educational web portal [4], it is required to find and study all the video lectures for this training course.

Let us also assume that the student found a video lecture in the WWW titled "Lecture 1. Physics of the atomic nucleus and particles." He suggests that this video lecture may be relevant to the training course being studied. Let us formulate a hypothesis.

Hypothesis 1. "Lecture 1. Physics of the atomic nucleus and particles" is taught by professor "I.M. Kapitonov" at the Faculty of Physics of the Moscow State University, and it is part of the training course titled "Physics of the atomic nucleus and particles."

To solve Task 1 and to verify the validity of Hypothesis 1, the following obvious reasoning should be performed step by step on the knowledge graph.

Step 1. On the educational web portal [4], from the drop-down list, select the knowledge graph “Nuclear Physics at MSU, MEPhI” (the fourth from the top in the list of knowledge graphs). Further, to solve Task 1 and to verify the validity of Hypothesis 1, one can start reasoning either with the corresponding classes “Training course,” “Training video,” “Professor,” etc., or with specific objects “Physics of the atomic nucleus and particles,” “Lecture 1. Physics of the atomic nucleus and particles,” “I.M. Kapitonov,” etc. Let it be decided to start the reasoning with the class “Training course.” One should type the first characters of the class name in the corresponding input field of the search widget, for example “Tr,” and then in the drop-down list, select the line “Training course.” To begin working with the knowledge graph, click the “Start” button, as shown in **Figure 2** below.

Step 2. Depending on the current setting of “Display of knowledge graphs” in a pop-up window, in a new tab of a web browser or in the same window, the workspace of the RDF browser and the corresponding graph node named “Training course” will be opened, see **Figure 3**. Each graph node has the form of a colored circle equipped with a button for displaying a local pop-up menu and a button for displaying the metadata. In addition, around each node there are petals of various sizes, shapes, and colors, with which it is possible to start step-by-step inductive

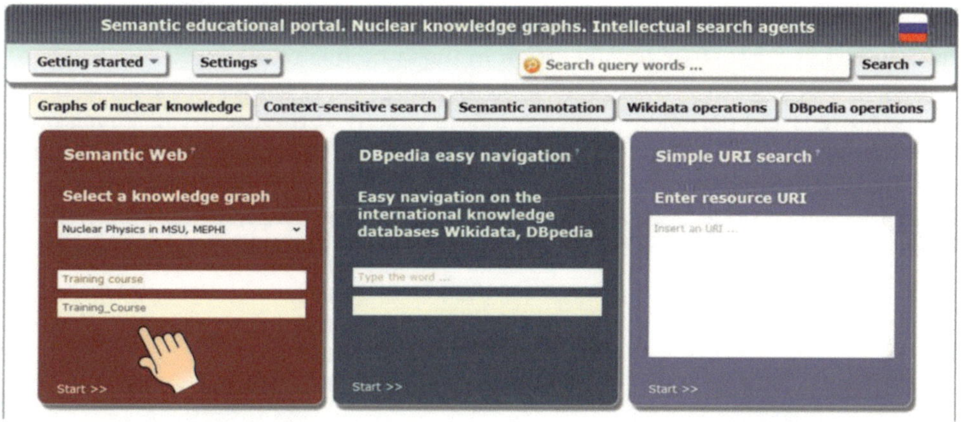

Figure 2.
Widgets for quick diving into the knowledge graphs: in the knowledge graph “Nuclear Physics at MSU, MEPhI” select the class “training course,” then click the “start” button.

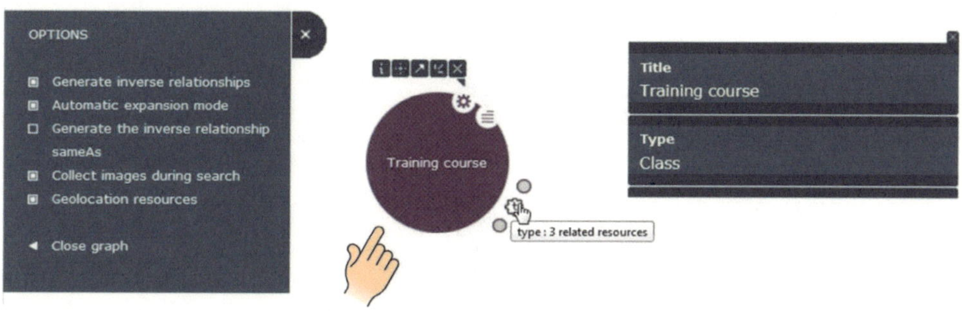

Figure 3.
RDF browser: the first node when diving into the knowledge graph “nuclear physics at MSU, MEPhI,” class “training course.”

reasoning and navigation in the graph. In the upper left part of the workspace of the RDF browser, there are options and help resources (legend, training videos, etc.). In the upper right part of the workspace of the RDF browser, the metadata associated with a particular node can be displayed if desired. Petals located around a node correspond to the single RDF triples in which this node is involved, or to the groups of such RDF triples (large petals). When user hovers over the petals, tooltips appear in which it is possible to see the names of the components of triplets. For large petals (triplet groups), the number of related resources is also displayed. Any group of triplets can be expanded or collapsed with a simple mouse click on the corresponding petal.

Step 3. We are interested in objects which have the type (i.e. belongs to the class) titled "Training course." There are three such objects and they are linked to our node by the "type" property, see **Figure 3**. Click to expand this resource group. Then go to the pop-up local menu of the "Training course" node and click the "View related resources" button (the second one on the right in the row of buttons). The RDF browser will display all the nodes associated with our node by any kind of properties, see **Figure 4**. Next, close the extra nodes and leave only those nodes that are associated with our node by the incoming "type" property, see **Figure 5**. In the course of practical work with the graph, it is advisable not to open the extra nodes by clicking only on the obviously necessary petals. The right side of **Figure 5** shows the metadata for the object named "Physics of the atomic nucleus and particles," which belongs to the class titled "Training Course." This object is an obvious candidate for further reasoning. However, **Figure 5** shows two other alternatives that can be left for further consideration in the inductive reasoning.

Step 4. The student is interested in the training course named "Physics of the atomic nucleus and particles," which is taught exactly at the Faculty of Physics of the Moscow State University. At this step, it is possible to narrow down the number of alternatives considered, taking an interest in the "teaches" property. **Figure 6** shows how this is done. Two alternative training courses taught at the National Research Nuclear University MEPhI, at this step, it is advisable to exclude from further considerations.

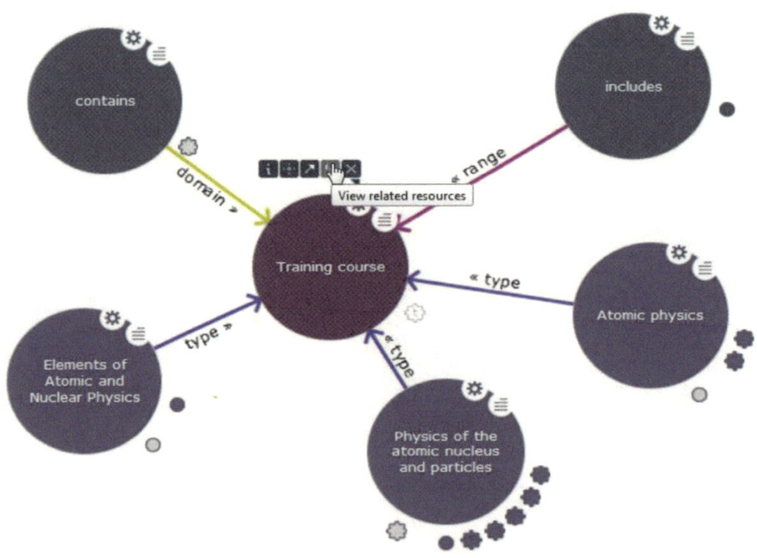

Figure 4.
RDF browser: displaying related resources for the class titled "training course."

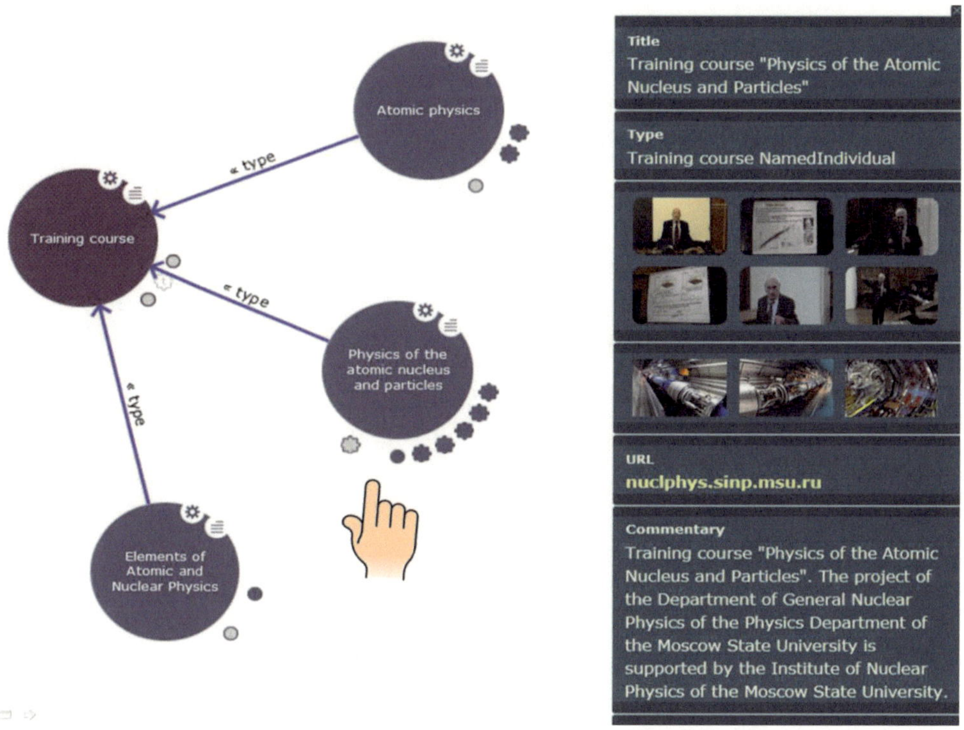

Figure 5.
RDF browser: Displaying the nodes of the graph, essential for continuing the reasoning, and the metadata for the object titled "physics of the atomic nucleus and particles."

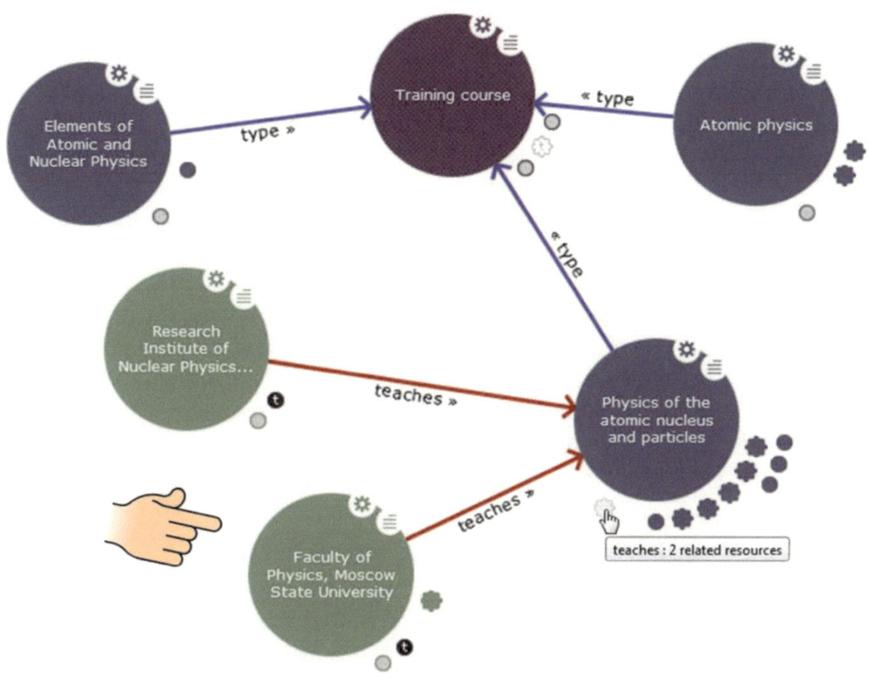

Figure 6.
RDF browser: Using the "teaches" property to reduce the number of alternatives under consideration.

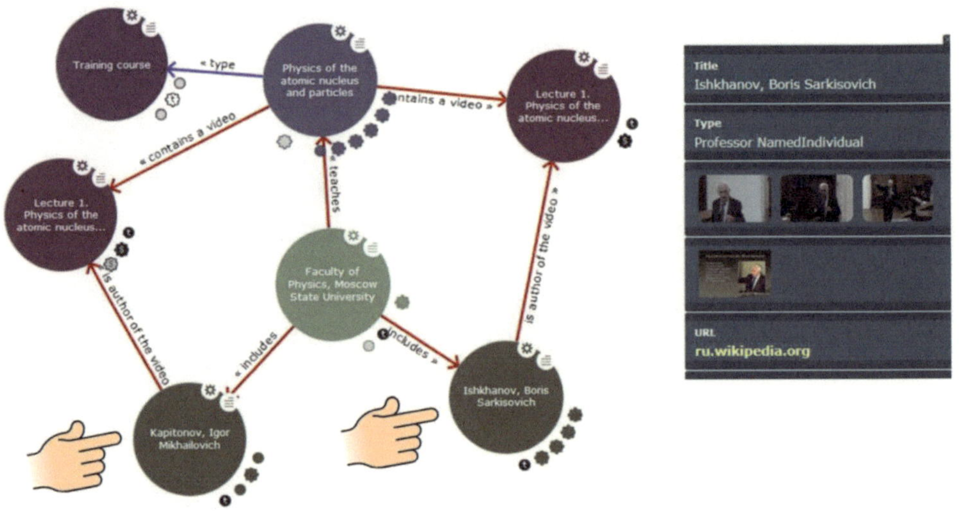

Figure 7.
RDF browser: Solving task 1 and confirming the validity of hypothesis 1.

Step 5. Continuing inductive reasoning for the object "Faculty of Physics, MSU" by the property "includes" and/or reasoning for the object "Physics of the atomic nucleus and particles" by the property "contains video," the student will be convinced of the validity of the Hypothesis 1 and will get the solution for Task 1, see **Figure** 7 below.

It is possible to view detected video lectures without leaving the workspace of the RDF browser, simply by clicking on the corresponding icon in the metadata area for the object named "Lecture 1. Physics of the atomic nucleus and particles."

The result obtained in Step 5 could be achieved in the course of deductive reasoning, without considering possible alternatives. However, the use of inductive reasoning allows the user to naturally extract additional knowledge from the graph, which will not be easy to obtain with a simple deductive inference [3].

Using the above method, it is easy to discover that professor "B.S. Ishkhanov" also gives lectures on the training course "Physics of the atomic nucleus and particles" at the Faculty of Physics of the Moscow State University, see **Figure** 7. All video lectures and other learning objects of both professors for this training course are available. Through the graph of knowledge, the full content of any training course and all the existing relationships are clearly revealed.

As can be seen from the above example, the inductive reasoning process in knowledge graphs resembles a computer adventure game, does not require special skills, and is accessible to an inexperienced user. Knowledge graphs similar to those considered are used in the real educational activity at the National Research Nuclear University MEPhI. Practice shows that university students master the methods of interactive work with knowledge graphs within a few minutes.

4. Knowledge acquisition: context-sensitive search

As a toolkit that prepares data for ontology refinement and enrichment, a software agent (which is essentially a specialized meta-search engine) for the reconnaissance context-sensitive search for new knowledge in the WWW is provided. To begin with it, several characteristic features of popular search engines that are well known to most users should be noted.

- The documents found are ranked by the public search engine in accordance with its internal algorithm, which does not always meet the interests of a particular user.

- Users are not always comfortable to manage the context of the search query, refine, and direct the search.

- Links to the commercial sites usually have a higher rating than other search results. Such effect is achieved through the use of the so-called search engine optimization (SEO) to artificially raise the positions of commercial network resources on pages of popular search engines, in order to increase the flow of potential customers for the subsequent monetization of traffic.

It seems that the above circumstances and trends make public search engines an increasingly inadequate tool for extracting knowledge in the WWW for educational purposes. The context-sensitive search is based on a simple idea: to create such an intermediary (software agent) between the knowledge engineer and public search engines, which helps to systematize search results in accordance with his professional needs, by effectively filtering inappropriate content and information garbage. The goal is to involve the power of the modern search engines in the maximum level, including built-in query languages and other search controls.

When the "Context-sensitive search" software agent is working, the global document search, as well as the search for specialized web resources, is initially performed by the regular search engines (Google Ajax Search, Yandex, Yahoo, and Mail.ru), the interaction with which occurs asynchronously via the dynamic pool of the proxy servers, each of which is hosted on the Google Cloud Platform. The results of the work of the regular search engines are a kind of "raw material" for further processing. Specially designed proxy servers on the cloud platform parse these results and generate the feeds, which are then sent to the client computer, where from the feeds, snippets are formed. These snippets, which contain metadata, before they appear on the monitor of the client computer, undergo additional processing, screening, and sorting, as described below. In particular, for each snippet, its relevance, persistence, and a number of other indexes are calculated, which are further used to systematize and clustering search results obtained.

5. Search context

The query language of some search engines may include the so-called "search context." It is about the use directly in the text of the search query of special operators, which allow the user to specify the presence and relative location of specific tokens in the documents found. In this paper, a "search context" is understood slightly different way, namely, as a certain restricted on length text that characterizes the domain that is currently of interest to the knowledge engineer.

When setting the search context, the following data sources are available: taxonomies, thesauri, keywords, ontologies, textual files from the client computer, and arbitrary resources from the WWW. Any combination of the above methods for setting the search context is allowed. The resulting context is the union of the selected options. The context defined in this way allows to select, sort, and organize information that comes from the search engines through the proxy servers.

Figure 8 below shows the possible options for setting the search context.

Figure 8.
Setting the context for the reconnaissance context-sensitive search: 1 – setting the context using a file from the client computer; 2 – setting the context using an arbitrary site; 3 – widgets to show the established context.

6. Relevance, pertinence, and metrics

For the purposes of this paper, the relevance of the snippet is the measure of the similarity between the snippet and the search query text. Under the pertinence of the snippet is meant the measure of the similarity between the snippet and search context that was defined earlier. These and other measures are calculated by means a fuzzy comparison of the corresponding texts. To quantify these measures, "Context–sensitive search" software agent uses the Levenshtein metrics [9]. The algorithm for calculating the relevance of the one particular snippet is as follows.

Each lexical unit (token) from the snippet is sequentially compared with each token from the text of the search query. In the case of an exact match of tokens, the relevance of a snippet is increased by number 3. If a complete match of the lexemes requires the use of one of the Levenshtein operations (insertion, deletion, and substitution of one symbol), then the relevance of a snippet is increased by number 2 and not 3 (2 = 3–1). Here, number 1 is the price of one Levenshtein operation. If a complete match of the lexemes requires the use of two Levenshtein operations, then the relevance of a snippet is increased by number 1 (1 = 3–2). Here, number 2 is the price of the two Levenshtein operations. In the case that more than two Levenshtein operations are required to match the lexemes, the relevance of the snippet does not increase at all. It is possible to finetune the prices (weights) of Levenshtein operations of each kind, which initially (by default) are all equal to one.

The algorithm for calculating the snippet's pertinence looks similar, with the only difference that each token from the snippet is successively compared to each token from the search context. As can be seen from the above description of the algorithm, the process of calculating the relevance and pertinence of snippets is a formal one, without analyzing the possible connections of individual tokens and their environment. It is assumed that earlier such an analysis was implemented to some extent during the initial search of documents and their full-text indexing in databases of regular search engines.

Various options for sorting search results in the final output of the "Context-sensitive search" software agent are allowed. The sorting by aspect named "dominance index" deserves a special mention, which provides a joint account of the values of many metrics that characterize the adequacy of the snippets. For example, the dominance index, in addition to the relevance and pertinence of the snippets, can also take into account the measure of the similarity between the snippet and the keywords, categories, and properties of the educational portal in total. For the practical calculation of the values of the dominance index, it seems reasonable to use the formalism of Pareto dominance relation [10], since Pareto's multicriteria ranking does not presuppose an a priori knowledge of the relative importance of aspects (for example, what is more important, relevance or pertinence?).

Let given the initial set of snippets, from which one should choose some optimal subset, the choice should be made on the basis of certain ideas about the adequacy of snippets (the principle of optimality). The selection task is a simple one, if there is only a single aspect by which it is possible to compare any two snippets and directly indicate which one is more adequate. The solution of the simple selection problems is obvious. In real situations, it is not possible to single out any one aspect. Moreover, it is often generally difficult to single out aspects. The selection and ranking of aspects that are essential for subsequent selection, in turn, is the task of choice. If some of the aspects are more important (priority) than other aspects, this circumstance should be taken into account in the mathematical model of choice.

The selection task is the algebra $<\Omega, \mathrm{O}>$ where Ω is a set of alternatives (in our case, a set of snippets) and O is the optimality principle. The task makes sense if the set of alternatives is known. Usually, the principle of optimality is unknown.

For further discussion, suppose that each snippet $x \in \Omega$ is characterized by a finite set of aspects $x = (x_1, x_2, \dots, x_m)$. Let $\mathrm{A} = \{1, \dots, m\}$ be the set of aspect numbers to consider when choosing; $\{\mathrm{A}\}$ is the set of all subsets A.

It can be assumed that choosing between any two snippets x and y with only one of any aspect taken into account is a simple task. If this is not the case, the corresponding aspect can be decomposed and presented as a group of simpler aspects. For each pair of snippets (x, y), we define a family of functions $\alpha_j(x, y)$ as follows:

$$\alpha_j(x,y) = \left\{ \begin{array}{ll} 1, & \textit{if } x \textit{ exceed } y \textit{ in aspect } j \\ 0, & \textit{if } y \textit{ exceed } x \textit{ in aspect } j \end{array} \right\} \textit{where } j \in \mathrm{A}; \quad x, y \in \Omega; \tag{1}$$

If x and y are equal or not comparable in some aspect with the number j, then for such number j, the function $\alpha_j(x, y)$ is not defined. Let us form a set J of numbers of such aspects that x and y differ in these aspects

$$J = \{\, j : j \in \mathrm{A}; \ \ \alpha_j(x,y) \ \textit{is defined}\}, \quad J \in \{\mathrm{A}\}; \tag{2}$$

Next, we construct a metric that takes into account the number of aspects by which a particular snippet is inferior to all other snippets. Let there be two snippets $x, y \in \Omega$. Denote

$$d(y,x) = \sum_{j \in J} \alpha_j(y,x) \tag{3}$$

the number of aspects in which y is better than x. Then, the value

$$D_\Omega(x) = \max_{y \in \Omega} d(y,x) \tag{4}$$

is called the dominance index of x when presenting the Ω set. This value characterizes the number of aspects of the snippet x that are not the best in comparison with all other snippets available in the Ω set.

Let us define the function $C^D(\Omega)$ for selecting the best snippets as follows:

$$C^D(\Omega) = \left\{ x \in \Omega : D_\Omega(x) = \min_{z \in \Omega} D_\Omega(z) \right\} \tag{5}$$

Here, the value $D_\Omega = \min_{x \in \Omega} D_\Omega(x)$ is called the index of dominance of the whole Ω set. Snippets with a minimum value of the dominance index form the Pareto set. The Pareto set includes snippets that are the best with respect to all the considered aspects, including relevance and pertinence.

In the project [4], an intuitively more acceptable value is used as the index of dominance, equal to the difference between the number of aspects taken into account and the dominance index determined by the formula (Eq. (4)). Groups of snippets with the same value of the dominance index form clusters, which in the final output of the "Context–sensitive search" software agent are arranged in descending order of this index.

As an illustration of the previous computations in the next section **Figure 9** shows a variant of sorting snippets by dominance index. Snippets are sorted in descending order of the dominance index value when six metrics are taken into account, including snippets relevance and pertinence. When snippets are ordered by the value of the dominance index, within groups of elements with the same value of the dominance index (that is, within a cluster), the snippets are ordered by each of the metrics taken into account in the calculations. Other ways to organize and systematize the content found are available for any combination of metrics that characterize the adequacy of the snippets.

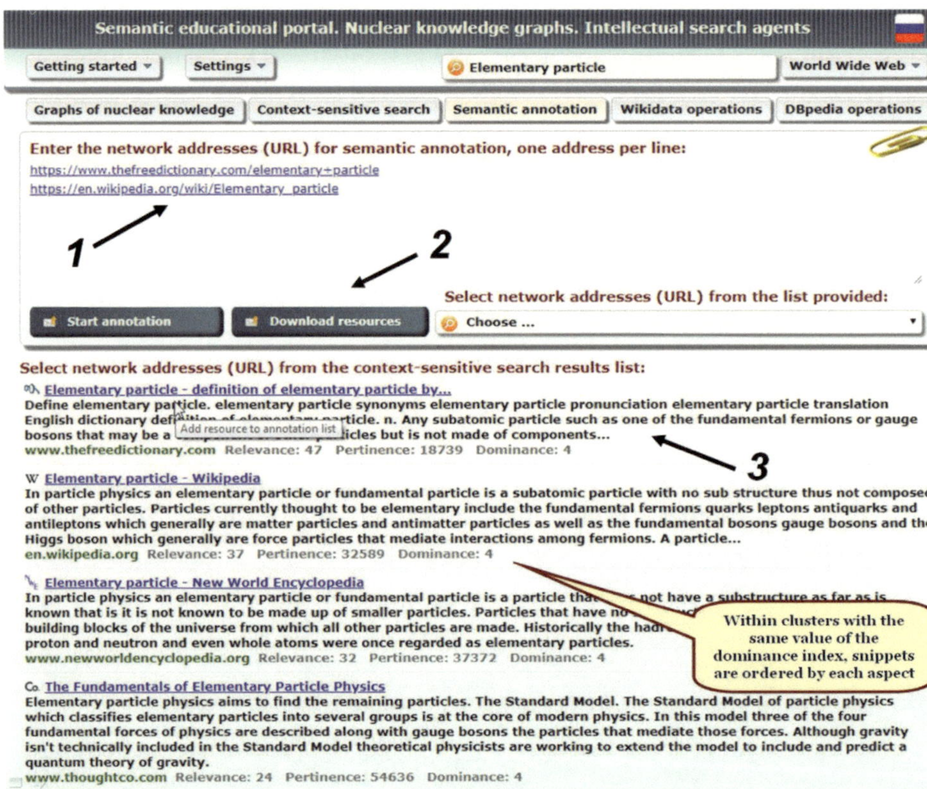

Figure 9.
Selecting network resources for semantic annotation: 1 – workspace for entering and editing network addresses (URLs) to be annotated; 2 – setting options and loading results of the context-sensitive search; 3 – the most relevant results of the context-sensitive search.

7. Knowledge graphs enrichment: semantic annotation

The database world is a place that is controlled by computers. Supercomputers have amazing computing capabilities, but they can be a struggle when it comes to acquiring new knowledge and experience or putting knowledge into practice. While it is easy for a human to decide whether two or more things are related based on cognitive associations, a computer often fails to do it. Unlike traditional lexical search where search engines look for literal matches of the query words and their variants, semantic annotation tries to interpret natural language close to how people do it. During semantic annotation, all references to cases related to entities in the ontology are recognized. Semantic annotation is the glue that ties ontologies into document spaces, via metadata.

The working panel for implementing the semantic annotation process is shown in **Figure 9** below. At the top of this panel is a workspace for entering and editing network resource addresses (URLs) to be annotated. The data in this workspace can be entered from any source, including manually. However, a more technologically advanced approach is to first find on the WWW those network resources that are most adequate to a given domain using the "Context–sensitive search" software agent. The found adequate content can then be easily loaded using the "Download resources" button and included in the list for annotation with a single mouse click.

The settings panel for the semantic annotation process is shown in **Figure 10** below. For annotation, you can select any of the knowledge graphs that are presented in the semantic repository, as well as any combination of them. To calculate measures of similarity between the annotated resource and entities from knowledge graphs, both text analysis methods and neural networks that are trained on existing knowledge graphs can be used.

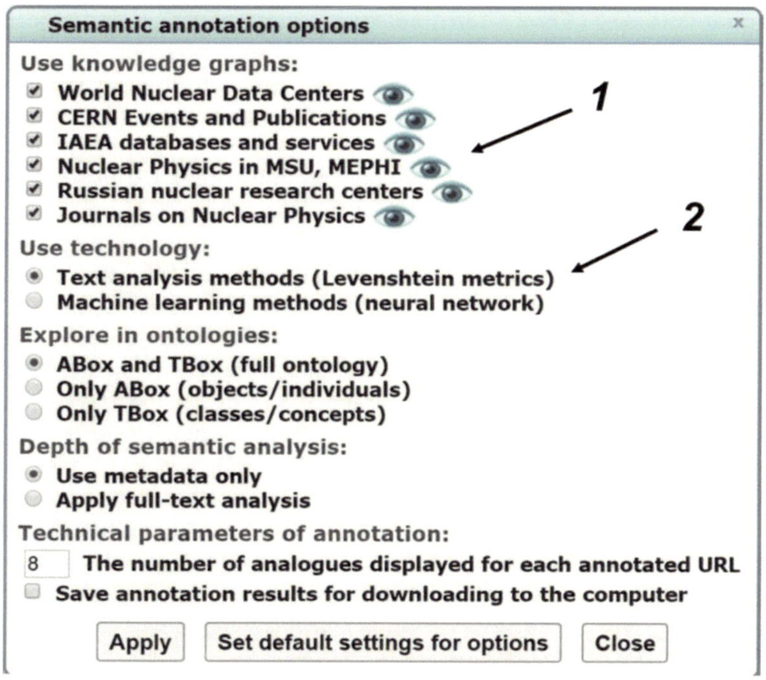

Figure 10.
Setting the options for the semantic annotation process: 1 – selecting and visualizing the knowledge graphs used; 2 – selecting of the technology and setting semantic annotation parameters.

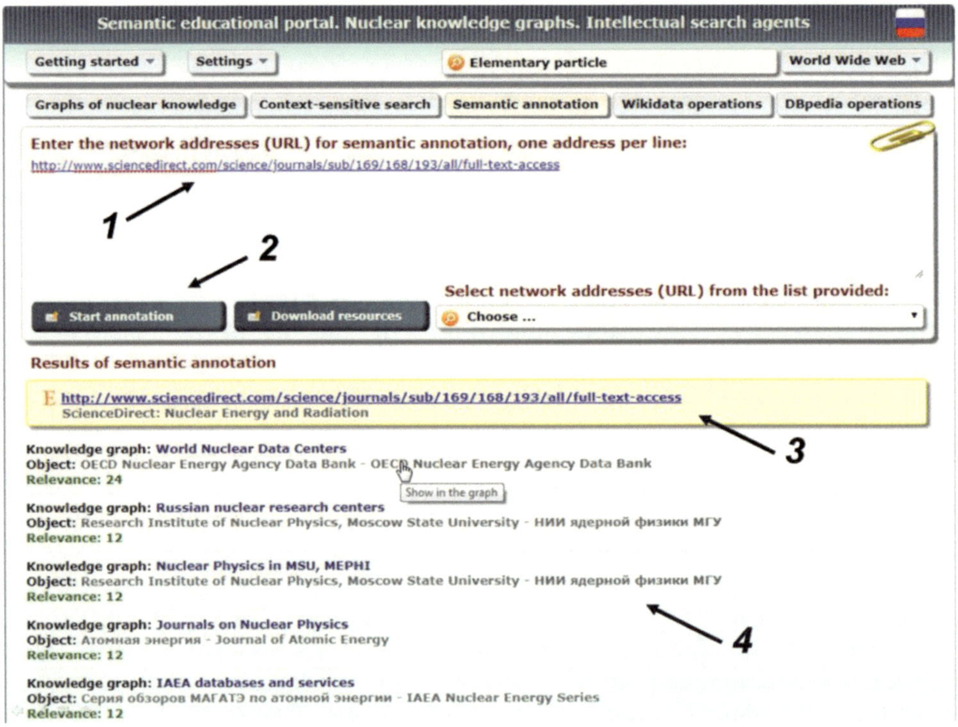

Figure 11.
Displaying semantic annotation results: 1 – addresses of the annotated network resources (URLs); 2 – setting options and starting the semantic annotation process; 3 – network resource for which semantic annotation is performed; 4 – knowledge graphs and entities corresponding to the annotated network resource.

It is possible to annotate network resources using classes (concepts) of the corresponding ontology (TBox – terminological components), using objects (individuals) of knowledge graphs (ABox – assertion components), or using both.

The depth of the carried out semantic analysis can be limited by considering only textual metadata inherent in network resources and entities in knowledge graphs. Full-text semantic analysis can be very expensive and, in many ways, redundant. Improving the accuracy of annotation in full-text analysis often does not justify the increased consumption of computing resources and time.

The number of displaying entities from knowledge graphs can be limited by the user. At the top of the output of the "Semantic annotation" software agent, the entities that are most adequate to the annotated resource appear. All the results of the work can be saved in files on the user's computer for later study.

As an example of using the "Semantic annotation" software agent, **Figure 11** below shows the results of the semantic annotation of one network resource. It can be seen that semantic annotations from five different knowledge graphs were discovered. With one click, the user can open the RDF browser and visualize the found annotations in any of the knowledge graphs, as well as anyone can see the surroundings of the entities found, for example, their classes and neighboring objects. This information is essential for a knowledge engineer who is engaged in knowledge graph refinement and enrichment.

8. Related work and conclusion

Groups of scientists from the University of Manchester, Stanford University, University of Bari and a number of other universities are focused on the issues of theory development and technology's implementation for semantic web, description logics and incarnations of the ontologies description language OWL. A recent qualified review [11] gives a fairly complete picture of the progress made in this area and the directions for further research.

Special mention should be made on the project [12], where for the first time an attempt was made to put into practice the methods of inductive reasoning for the purpose of semantic annotation of content from the WWW. As for the issues of visualization linked data [13], here, one of the first successful projects was Lodlive [14], which provided a tool for easier surfing through the DBpedia knowledge database. It is important to continue to develop and improve tools for intuitive perception of linked data for non-professionals. VOWL [15] is one of the modern project for the user-oriented representation of ontologies; it proposes the visual language, which is based on a set of graphical primitives and an abstract color scheme. As noted in [3], LinkDaViz [16] proposes a web-based implementation of workflow that guides users through the process of creating visualizations by automatically categorizing and binding data to visualization parameters. The approach is based on a heuristic analysis of the structure of the input data and a visualization model facilitating the binding between data and visualization options. SynopsViz [17] is a tool for scalable multilevel charting and visual exploration of very large RDF & Linked Data datasets. The adopted hierarchical model provides effective information abstraction and summarization. Also, it allows to efficiently perform the statistic computations, using aggregations over the hierarchy levels.

In contrast to the above solutions, the project [4] is mainly focused on the implementation in educational activities of universities and is not limited to visualization of knowledge graphs and interactive navigation, but is aimed at the introduction of the latest semantic web technologies to the training process, taking into account the achievements in the field of uncertain reasoning. The results obtained and the software created are used in the real educational process at National Research Nuclear University MEPhI, and the project, as a whole, is focused on the practical mastering of semantic web technologies by students and professors.

Acknowledgements

The reported study was funded by the Russian Foundation for Basic Research and Government of the Kaluga Region according to the research projects 19-47-400002 and was funded by the Vladimir Potanin Foundation according to the project GC190001383.

Author details

Victor Telnov* and Yuri Korovin
National Research Nuclear University MEPhI, Obninsk, Russian Federation

*Address all correspondence to: telnov@bk.ru

References

[1] W3C OWL 2 Web Ontology Language. 2012. Available from: http://www.w3.org/TR/owl2-overview/ [Accessed: 29 March 2020]

[2] Baader F, Calvanese D, McGuinness D, Nardi D, Patel-Schneider P. The Description Logic Handbook: Theory, Implementation and Applications. 2nd ed. New York: Cambridge University Press; 2010. p. 505

[3] Telnov V, Korovin Y. Semantic web and knowledge graphs as an educational technology of personnel training for nuclear power engineering. Nuclear Energy and Technology. 2019;**5**(3): 273-280. DOI: 10.3897/nucet.5.39226

[4] Telnov V. Semantic educational portal. Nuclear knowledge graphs. Intellectual search agents [Internet]. 2020. Available from: http://vt.obninsk.ru/x/ [Accessed: 29 March 2020]

[5] ISO 19505 UML Part 2 Superstructure. 2012. Available from: http://drive.google.com/file/d/0B0jk0QU2E5q9NVIwMFNieGxOZVU [Accessed: 29 March 2020]

[6] Ontology example "Nuclear Physics at MSU and MEPhI". 2020. Available from: http://drive.google.com/file/d/1AIXMsm3cfAxR6NX220R4ZeFeoSFp0mj5 [Accessed: 29 March 2020]

[7] Fanizzi N, d'Amato C, Esposito F. Induction of concepts in web ontologies through terminological decision trees. In: ECML/PKDD. Barcelona. Spain; 2010. pp. 442-457. DOI: 10.1007/978-3-642-15880-3_34

[8] Bobillo F, Carvalho R, Costa P, d'Amato C, Fanizzi N, Laskey K, et al. Uncertainty reasoning for the semantic web III. In: SWC International Workshops URSW. Revised Selected Papers; 21–25 October. Sydney, Australia; 2013. pp. 1-328. DOI: 10.1007/978-3-319-13413-0

[9] Levenshtein V. Binary codes capable of correcting deletions, insertions and reversals. Soviet Physics – Doklady. 1965;**10**(8):707-710

[10] Chen Y, Wang Z, Yang E, Li Y. Pareto-optimality solution recommendation using a multi–objective artificial wolf–pack algorithm. In: Proceedings of 10th International Conference on Software, Knowledge, Information Management & Applications (SKIMA); 15–17 December 2016. Chengdu, China; 2016. pp. 116-121. DOI: 10.1109/SKIMA.2016.7916207

[11] d'Amato C. Machine learning for the semantic web: Lessons learnt and next research directions. Semantic Web. 2020;**11**:195-203. DOI: 10.3233/sw-200388

[12] d'Amato C, Fanizzi N, Fazzinga B, Gottlob G, Lukasiewicz T. Combining Semantic Web Search with the Power of Inductive Reasoning. 2013. Available from: http://ceur-ws.org/Vol-527/paper2.pdf [Accessed: 29 March 2020]

[13] Bikakis N, Sellis T. Exploration and Visualization in the Web of Big Linked Data: A Survey of the State of the Art. 2016. Available from: http://arxiv.org/pdf/1601.08059.pdf [Accessed: 29 March 2020]

[14] Camarda D, Mazzini S, Antonuccio A. Lodlive, exploring the web of data. In: Proceedings of the 8th International Conference on Semantic Systems, I-SEMANTICS, ACM; September 2012. Graz, Austria; 2012. pp. 197-200

[15] Schlobach S, Janowicz K. Visualizing ontologies with VOWL. Semantic Web. 2016;7:399-419. DOI: 10.3233/SW-150200

[16] Thellmann K, Galkin M, Orlandi F, Auer S. LinkDaViz – Automatic binding of linked data to visualizations. In: Proceedings of the 15th International Semantic Web Conference; October 2015. Bethlehem, PA, USA; 2015. pp. 147-162

[17] Bikakis N, Skourla M, Papastefanatos G. Rdf:SynopsViz - a framework for hierarchical linked data visual exploration and analysis. In: Proceedings of the European Semantic Web Conference ESWC; May 2014. Heraklion, Crete, Greece; 2014. pp. 292-297